Courtesy of

DE BEERS

GROUP OF COMPANIES

ISBN 978-99912-941-8-6


Published by Michael C. Brook, 2012
Kwena Pools, Notwane,
P O Box 25283, Gaborone, Botswana

EDITOR
Esther Kanaimba-Senai

AUTHOR'S ADDRESS
Written by Michael C. Brook,
Kwena Pools, Notwane,
P O Box 25283, Gaborone, Botswana
email: mcbrook@hotmail.com

LAYOUT
Adobe InDesign

DESIGN AND LAYOUT
Charles M. Njanjari
email: charliev75@yahoo.co.uk

PRINTED BY
Longpack Co., Ltd.
805A, No.8-9, Max Mall
1500 Lane Lianhua South Rd.
Shanghai 201108 PR China
email: info@longpack.com

# Sponsors

The author would like to thank the following companies and organisations for their support in the publication of this book:-

DEBSWANA
Mining the resource, enriching the nation

MMEWR
MINISTRY OF MINERALS, ENERGY AND WATER RESOURCES

DTC Botswana

DE BEERS
A DIAMOND IS FOREVER

Barloworld Equipment CAT

SHRENUJ
HERITAGE · INNOVATION · PASSION

DWDF
DEWET DRILLING BOTSWANA

ROSCOE MOSS COMPANY

BAUER DEWET

ABN·AMRO

PetraDiamonds
Botswana

Steinmetz
*Creators of the world's finest diamonds*

Private collection

# BOTSWANA'S DIAMONDS

## Prospecting to Jewellery

# TABLE OF CONTENTS

*The Aquarium diamond recovery plant at the world's richest diamond mine - Jwaneng Mine, Botswana (Debswana)*

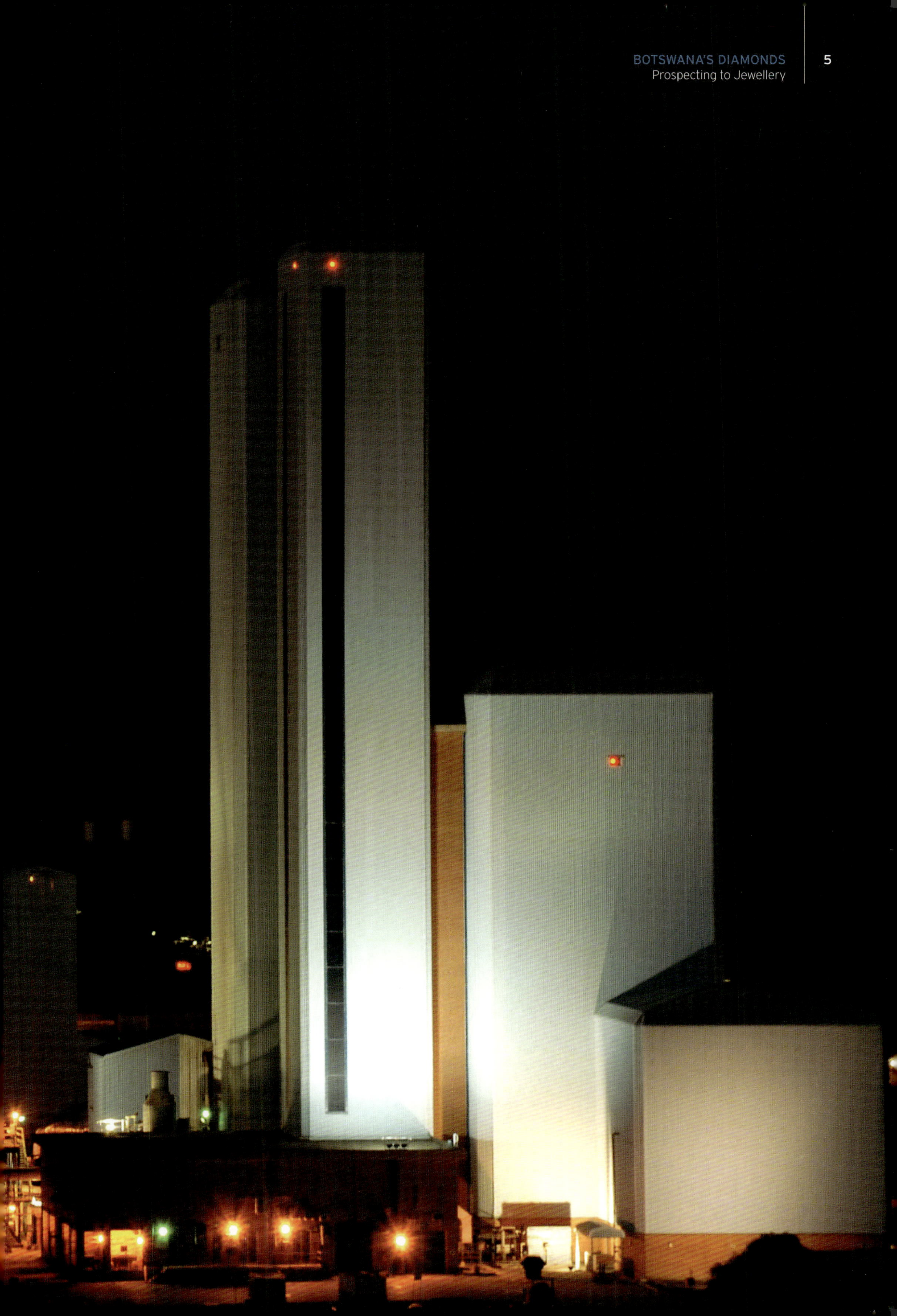

# Foreword

By Hon. Dr. Ponatshego Kedikilwe
Vice President of the Republic of Botswana and
Minister of Minerals, Energy and Water Resources

The discovery of the first economic deposit of diamonds in Botswana, located in the Orapa Kimberlite field in the Central District happened in 1967, just one year after Botswana gained independence from Britain. Today, Botswana is the world's largest producer of diamonds and there are currently seven operating mines, with an eighth, Botswana's first underground mine, due to commence production in 2013.

Botswana's good governance since 1966, when it became independent, has allowed the significant diamond revenue accrued by Government to be optimally used to develop a nation, which was once one of the poorest in the world to the middle-income economy of today with one of the largest Gross Domestic Products on the African Continent. Revenue from Botswana's diamond mines currently contributes about 35% of the nation's Gross Domestic Product and over 60% of revenue to the Botswana Government. Revenue from diamond sales has spearheaded Botswana's economic growth and development.

Diamond prospecting was an enormous challenge in the remoteness of the Kalahari Desert which makes up more than 80% of the country. In fact, it took one and a half decades after the first few rough diamonds were discovered in 1955 to open Botswana's first diamond mine at Orapa in 1971.

Today, Orapa operates as the most productive mine in the country and was joined by three other Debswana operations at Letlhakane (1977), Jwaneng (1982) and Damtshaa (2003). Jwaneng is the world's richest diamond mine. In addition, Lerala Mine was opened in 2008 and B/K11 and Karowe Mines in 2011. The Ghaghoo Mine is scheduled to come on line in 2013.

In this book, "Botswana's Diamonds; Prospecting to Jewellery", we can read, for the first time, the truly remarkable history of the discovery of diamonds in Botswana and the different components that make up the Botswana Diamond Pipeline today, being the supply chain of diamonds, from prospecting to mining, to diamond recovery, to rough diamond sorting, valuation, sales and marketing, to diamond polishing and cutting, and finally to diamond jewellery manufacturing. In Botswana, we can now witness the route of the diamond from "Mine to Store" or from "Rough to Finger".

The last few years has seen tremendous growth in the beneficiation of Botswana's diamond industry, involving the expanded development of the downstream stages of the pipeline, thus benefitting Botswana and Batswana substantially.

In 2008, the Botswana Government clustered a number of major development projects into six hubs to attract internal and external investment. A Diamond Hub was established to facilitate beneficiation and promote Botswana as one of the world's major diamond trading centers. The following additional initiatives have also been supported by Government:-

- The formation of the Botswana Diamond Manufacturer's Association in 2007

- The construction of a Diamond District incorporating a Diamond Technology Park in 2008

- The 2008 construction of a Diamond Trading Company Botswana sort house, the largest sorting and valuation facility in the world

- A strategy for the development of diamond cutting, polishing and jewellery making skills, launched in 2009

- The construction of a Secure Transfer Facility (STF) at Sir Seretse Khama International Airport in 2011

- The 2011 agreement by DTC International to move all of its sales and other operations to Botswana before the end of 2013

- The formation in early 2012 of the De Beers Aggregation Company in Botswana which will undertake the aggregation of all De Beers worldwide diamond production in Botswana for the first time

- The formation of Botswana's first state diamond trading company, the "Okavango Diamond Trading Company", in 2012, which will sell diamonds independently from DTCB, commencing with 10% of the run of Debswana's total annual production

- The construction of the state diamond company's diamond auction house, scheduled for completion in 2012

It is my honest belief that Botswana is now ready to become one of the leading Diamond Centres in the world and it is my dream and desire that it eventually develops into the world's largest Diamond Hub.

**Hon. Dr. Ponatshego Kedikilwe**
Vice President of the Republic of Botswana and Minister of Minerals, Energy and Water Resources

# Geographical Location of Botswana

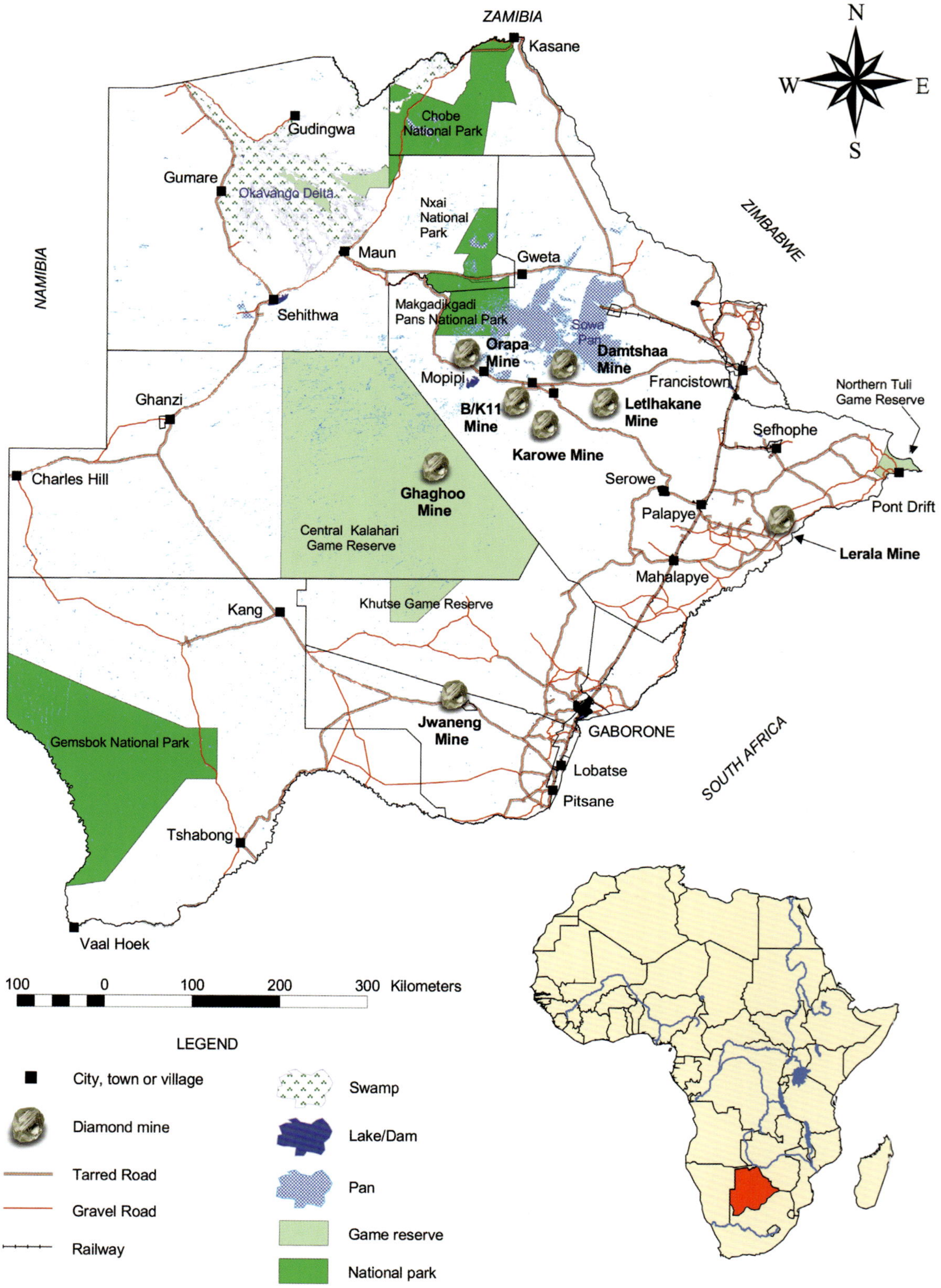

# Introduction

Botswana is the world's largest diamond producer. It is also the largest diamond producer by value.

The first diamondiferous Kimberlite (the host volcanic rock which brings diamonds to the earth's surface), known as B/K1 pipe, was discovered in Botswana 15km north of Letlhakane in 1967, a year after Botswana gained independence from British colonial rule over the then Bechuanaland. However, the first recognized diamond find in Botswana, in alluvial deposits, was made by Consolidated African Selection Trust prospecting geologists in 1955 on the banks of the Motloutse River at Foley siding, some 250 km east of Orapa Town.

Diamonds, however, had been found in India 4,000 years ago, but their true worth and preciousness was only realized at the commencement of the gold and diamond rush in South Africa in the 1880s. Today, Southern and Central Africa remain the most productive gem quality diamond areas in the world, followed by Russia, then Canada.

The first recorded diamond find in Southern Africa was in 1866 on the banks of the Orange River at the border of Namibia and South Africa; the "Eureka Diamond or Star of South Africa", weighing a mighty 83.5 carats (one carat being equivalent to 0.2 grams), which sparked a diamond rush. Between 1869 and 1871 six diamond pipes began to be mined in South Africa, including the now "Big Hole" at Kimberly. Cecil John Rhodes arrived in South Africa from England at the "New Rush" settlement, renamed Kimberly in 1873 and subsequently the igneous rock formation, Kimberlite, was named after this town. The De Beers Mining Company Ltd, the forerunner of the De Beers Family of Companies today, was founded in 1880 by Cecil John Rhodes, Charles Rudd and other partners and in1888, De Beers Consolidated Mines (DBCM) was established.

On 30th June, 1932, DBCM wrote to the British High Commissioner in Pretoria, South Africa, in order to acquire a Crown Grant in respect of diamond prospecting and mining over the area of the Bechuanaland Protectorate not already held under concession by the British South Africa Company. Pretoria then wrote to No 10 Downing Street (the office of the British Prime Minister), England, who replied that the application could not be processed because the grant would be subject to a Mining Law still being drafted. The actual first draft of several mining proclamations was released on 14th October, 1931, under the Bechuanaland High Commissioner's Office, Notice No. 127 of 1931.

Several years later, a letter from DBCM to the Chief Mining Commissioner, Bechuanaland Protectorate, whose capital was then in Mafikeng, dated 16th May 1938, admitted failing to take out a renewal of licence No 705, and begged the Commissioner to renew for the usual sum then of One British Pound.

# Introduction

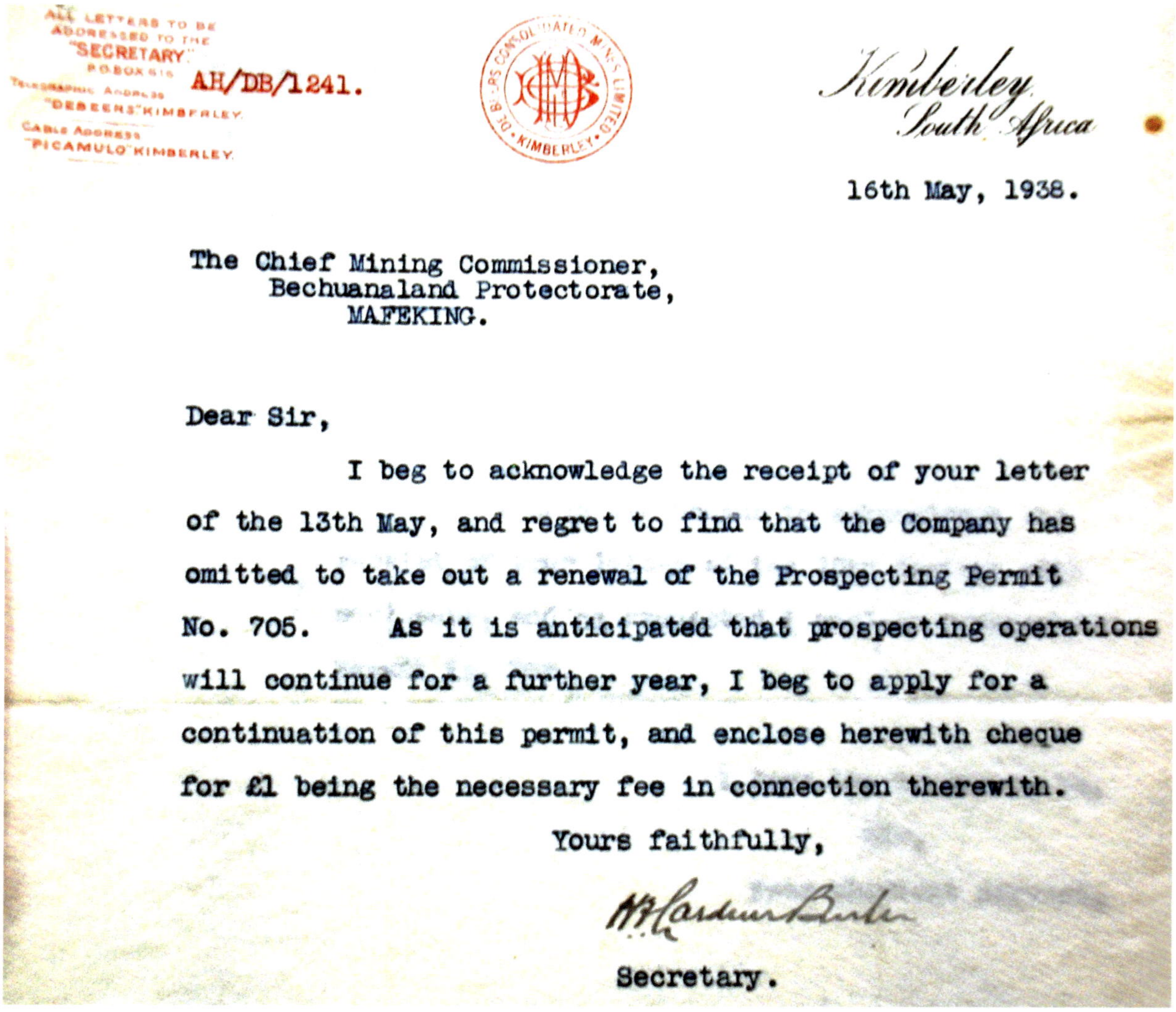
ALL LETTERS TO BE ADDRESSED TO THE "SECRETARY" P.O.BOX 616
TELEGRAPHIC ADDRESS "DEBEERS" KIMBERLEY.
CABLE ADDRESS "PICAMULO" KIMBERLEY.

AH/DB/1241.

Kimberley, South Africa

16th May, 1938.

The Chief Mining Commissioner,
Bechuanaland Protectorate,
MAFEKING.

Dear Sir,

I beg to acknowledge the receipt of your letter of the 13th May, and regret to find that the Company has omitted to take out a renewal of the Prospecting Permit No. 705. As it is anticipated that prospecting operations will continue for a further year, I beg to apply for a continuation of this permit, and enclose herewith cheque for £1 being the necessary fee in connection therewith.

Yours faithfully,

Secretary.

*1938 application by De Beers Consolidated Mines (DBCM) to renew a diamond prospecting permit (Botswana National Archives and Records)*

Permission was granted, and licences were renewed for many years until diamond prospecting began in earnest only in 1955, in the name of a DBCM company called Kimberlitic Searches.

A very important point to make early on is just how rare and therefore valuable diamonds are. Of the approximately 6,500 kimberlites known globally to date (2010), less than 3% are diamondiferous and less than half of those are economically viable to be mined (de Wit, 2010). More than 380 kimberlites have been found in Botswana and there are currently seven diamond mines in the country producing almost 40% of the world's total diamond production. Most of the other diamond production comes from mines elsewhere on the African Continent and from Russia, Australia, Canada and South America. In Africa, the other diamond producing countries are Angola, Central African Republic, Democratic Republic of Congo, Ghana, Guinea, Ivory Coast, Lesotho, Liberia, Namibia, Sierra Leone, South Africa, Tanzania and Zimbabwe. Kimberlites have also been discovered in Gabon, Kenya, Mali, Mauritania, Swaziland, Uganda and Zambia.

*Seretse Khama being sworn in as First President of Botswana after his party Botswana Democratic Party (BDP) won the 1979 elections (BOPA)*

On 30th September 1966, Sir Sereste Khama was sworn in as Botswana's first President. To date, Botswana has had only four Presidents, a testimony to good governance which has provided a stable and reliable political platform upon which to build Botswana's well deserved prosperity.

Indeed, a year later in 1967, the same year the very first diamondiferous kimberlite was found in Botswana, all the tribal chiefs of Botswana signed an agreement with President Seretse Khama, surrendering all mineral rights in the tribal areas to the Botswana Government so that the prospective wealth of minerals could be shared among all their districts in Botswana.

In the words of Dr. Gaositwe K.T Chiepe, former Minister of Mines and Natural Resources, Trade and Industry, Foreign Affairs and Education and also the first female cabinet member in Botswana (1974), "good governance is essential to ensure that the benefits of diamonds can be spread as evenly as possible".

*Dr. Gaositwe K.T Chiepe, former Minister of Mines and Natural Resources (Botswana National Archives and Records)*

Dr. Chiepe was Minister of Mines and Water Affairs between 1977 - 1984 when both Orapa and Jwaneng diamond mines were opened and she took a leading role in the negotiations between the Botswana Government and De Beers.

She said *"the talks were tough but fair"* and that *"it was fantastic to feel that I was helping to shape the future of my country. Today diamond mines are the backbone of our economy, providing the people with clinics, schools, roads, employment and new skills, and they finance drilling all over the country for water – our scarcest resource".*

# Introduction

*Botswana Chiefs meet in Gaborone in 1967 to agree to accept relinquishing of mineral rights in their tribal lands to the National Government. Back row, L-R, Minister of Commerce, Industry and Water Affairs, Mr. Haskins, Messrs O. Porogwe, A. Matsietsa (Bamalete), S. Matlhabaphiri, N.B. Kgosie ntsho (Bakwena), B. Pilane, C. Ratsheko (Bakgatla), and the Minister of Local Government, Mr. Kgabo. Front row, L-R, Regent Kelemogile Mokgosi, Chief Neale Sechele, Mr. Masire and Chief Linchwe. (Botswana National Archives and Records, 1967)*

# Introduction

## Botswana Diamond Timeline

### The Seven Diamond "Periods" in Botswana

| DATE | PERIOD | ACTIVITY |
|---|---|---|
| 1954 - 1966 | The Lean Mean period | Preparing to prospect, rudimentary methods, poor success rates |
| 1967 - 1974 | The Wonderful Discovery era | Numerous diamondiferous kimberlites discovered – all of today's major mines discovered during this incredibly successful prospecting period |
| 1971 - 1982 | A decade of Mine openings | Orapa, Letlhakane and Jwaneng Mines opened |
| 1980 - Present | Prospects re-visited | Old De Beers prospects re-visited by newcomers who achieved success through improved technology and methods |
| 2003 - Present | The smaller Mines | Damtshaa, Lerala, B/K11, A/K6 opened and Ghaghoo under construction |
| 2006 - Present | Valuation and Trading Consolidation | Improvement and consolidation of Diamond sorting, valuation and retailing |
| 2008 - Present | Towards Greater Diamond Beneficiation | The new road to added value from downstream activities – polishing, cutting, Jewellery and sales |

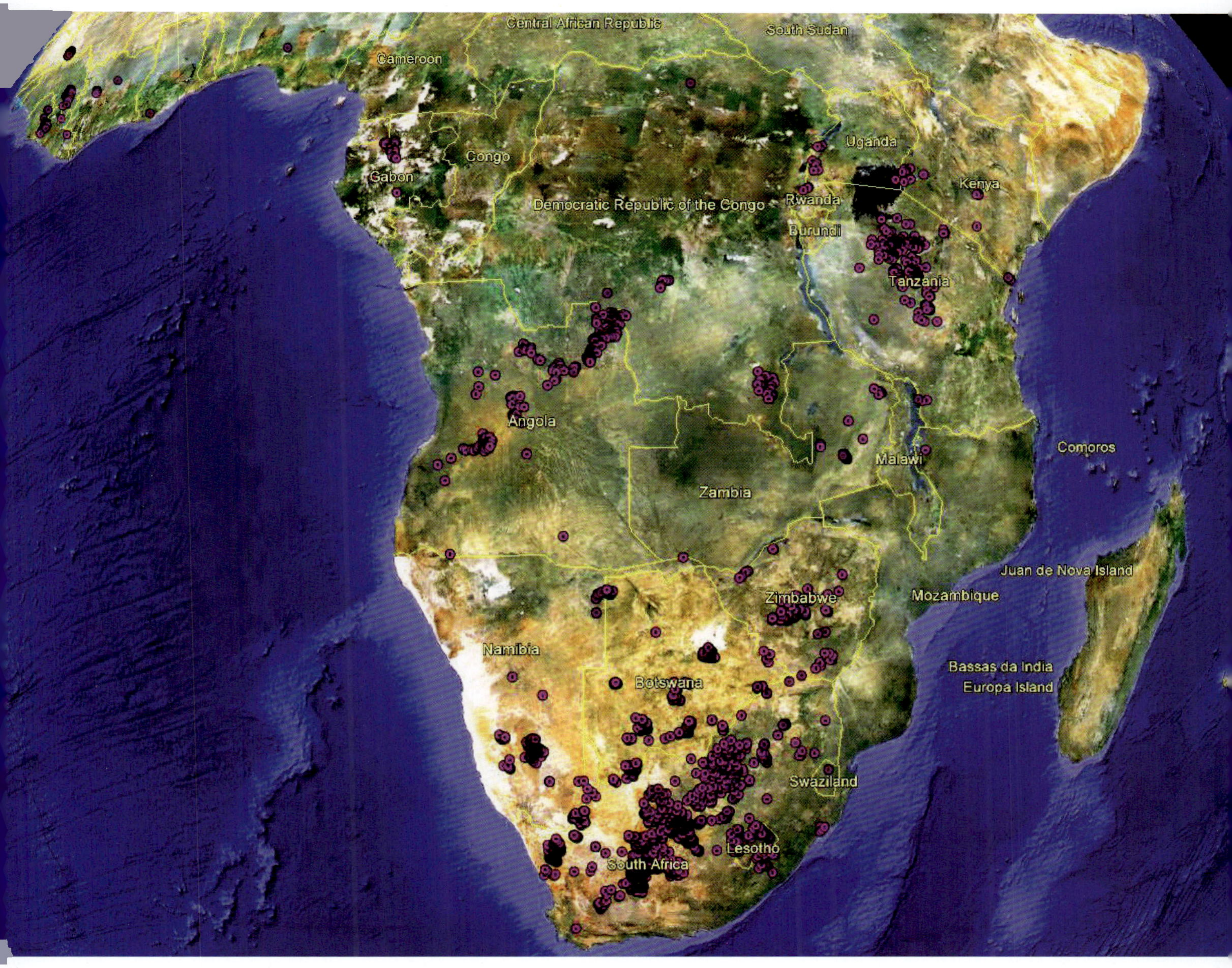


*Location of kimberlites on the African Continent (Google Earth, World Wide Kimberlite Database 2008)*

# Introduction

## Botswana **Diamond Timeline**

### 1938

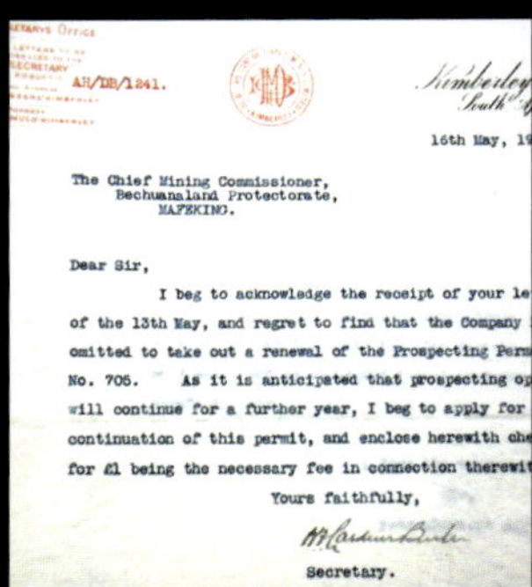
AN/2B/1241.

16th May, 19

The Chief Mining Commissioner,
Bechuanaland Protectorate,
MAFEKING.

Dear Sir,

I beg to acknowledge the receipt of your le
of the 13th May, and regret to find that the Company
omitted to take out a renewal of the Prospecting Perm
No. 705. As it is anticipated that prospecting op
will continue for a further year, I beg to apply for
continuation of this permit, and enclose herewith ch
for £1 being the necessary fee in connection therewit

Yours faithfully,

Secretary.

Kimberlitic Searches (De Beers Consolidated Mines-DBCM) applies for first Crown Prospecting Grants in Bechuanaland

### 1954

DBCM program of diamond prospecting commences in Bechuanaland, starting in Tuli Block lead by Dr. Gavin Lamont

### 1955

First diamonds found by Consolidated African Selection Trust (CAST) Prospecting Geologists at Foley near Motloutse River

### 1960

DBCM commence prospecting in Ghanzi Farms Block

### 1962

DBCM issued Crown Grant over Bamangwato tribal territory to follow up CAST diamond find

### 1964

Dr. Gavin Lamont applies Dr. Alex du Toit's crustal warping theory to suggest Foley diamonds came from a primary source in the Kalahari Desert to the west

### 1966

First two Botswana primary para – "Kimberlites" (non-diamondiferous lamprophyres) discovered by DBCM at Morobe and Ramokolonyane near Mochudi

On 30th September, Sir Sereste Khama was sworn in as Botswana's first President. Botswana replaces the Bechuanaland Protectorate under British Rule

### 1966

December, Dr. Louis Murray extends DBCM prospecting life in Botswana after Dr. Lamont and Jim Gibson's positive indicator finds in the Orapa – Letlhakane area from their rapid winter reconnaissance prospecting trip

### 1967

Botswana Dikgosi sign agreement with the President relinquishing their individual tribal rights to precious stones discoveries

1st March, Manfred Marx discovers Botswana's first diamondiferous kimberlite, 2125 B/K1 15km north of Letlhakane village

17th March, second diamondiferous kimberlite, 2125 B/K2 discovered in Orapa – Letlhakane area

25th April, kimberlite 2125 A/K1 discovered at Orapa ( later to become Orapa Diamond Mine)

### 1968-1972

Four Damtshaa pipes discovered by De Beers (B/K9 &B/K12 to become Damtshaa Mine in 2003)

**1969**

Letlhakane D/K1 &D/K2 discovered in June (to become Letlhakane Mine in 1977)

23rd June, De Beers Botswana Mining Company (Pty) Ltd - 85% De Beers, 15% Botswana Government formed, subsequently renamed Debswana in 1992

First batch of Debswana sponsored students sent to London under what later became a long lived scholarship program

**1970**

A/K6 Pipe (now Karowe Mine) operated by Boteti Mining, discovered by De Beers

**1971**

March, Discovery of diamondiferous kimberlite pipes D/K2 (later to become Jwaneng Mine) and D/K7 at Jwaneng

1st July, Orapa Mine Officially opened by Sir Seretse Khama

BDVC; Botswana Diamond Valuing Company (BDVC) formed – Botswana Government: De Beers 50:50 joint venture

**1973**

Debswana Main Mall Head Office constructed

**1974**

B/K11 (now mine operated by Firestone) discovered by De Beers

**1975**

De Beers Botswana Mining Company became a 50:50 partnership between the Botswana Government and De Beers

Construction of Letlakhane Mine commenced

**1976**

Diamond Manufacturing Botswana (DMB) becomes first diamond polishing factory in Botswana

**1977**

Commencement of production from Letlhakane Mine

**1978-1982**

Falconbridge discover 62 kimberlites in four fields, Tsabong and Kokong Fields discovered in 1978

**1980**

July 13th, Seretse Khama, first President of Botswana dies and Dr. Quett Masire replaces him as second President of Botswana

Gope 25 kimberlite (now Ghaghoo Mine) discovered by Falconbridge

**1982**

14th August, Jwaneng Mine officially opened by President Ketumile Masire

Botswana Diamond Valuing Company Sorting building, "Orapa House" opened in Gaborone

**1987**

Botswana Government acquires 5% interest in De Beers

**1989**

Jwaneng re-crush plant commissioned

## 1991-1992

De Beers discover Lerala kimberlites (to be opened by DiamonEX as Lerala Mine in 2008)

## 1992

On 25th March, De Beers Botswana Diamond Mining Company renamed Debswana

## 1996

Debswana agrees to double production at Orapa

## 1997

24hour Mining Operations commence at Debswana

## 1998

Festus Mogae sworn in as third President of Botswana on 1st of April

## 2000

Commissioning of Jwaneng Mine Aquarium (a first of its kind globally)

Orapa 2000 Expansion- Official Opening of Number 2 Plant by President Festus Mogae and Nicky Oppenheimer, Chairman of De Beers

## 2001

Botswana Government increases its stake in De Beers from 5% to 15%

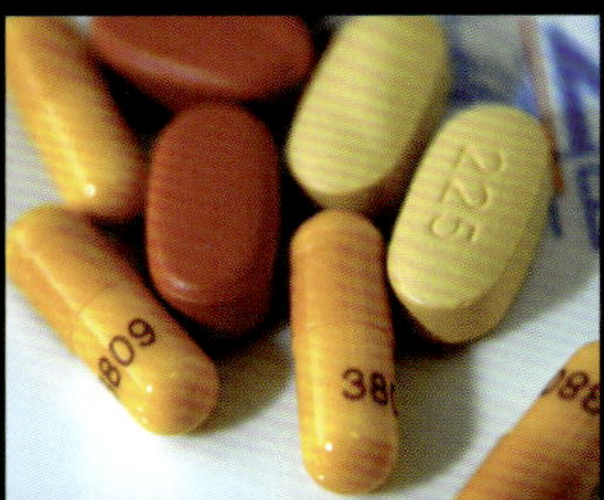


Debswana becomes the first company in the world to provide free Anti Retroviral Treatment to its employees

## 2002

Damtshaa Mine is commissioned

## 2003

Damtshaa Mine officially opened by Festus Mogae on October 25th

## 2004

Debswana successfully negotiates with the Botswana Government new 25-year leases for all its four diamond mines. The new 25-year mining licences will run until 2029

## 2006

Government and De Beers sign agreements to establish Diamond Trading Company Botswana (DTCB) - 50:50 joint venture between Botswana Government and De Beers. Left to right, Dr Akolang Tombale, Charles Tibone and Nicky Opppenheimer

DTC Botswana

## 2007

Debswana signs a new 25 year lease agreement with Ngwaketse Land Board for the Jwaneng Diamond mine.

February, DTCB Diamond Academy launched

DTCB signs three year supply contracts with 16 cutting factories ("Sightholders") licenced in Botswana.

- Dalumi Botswana (Proprietary) Limited
- Zebra Diamonds (Proprietary) Limited
- Yerushalmi Bros Botswana (Proprietary) Limited
- Teemane Manufacturing Company (Proprietary) Limited (Parent company Diarough NV)
- Suashish Diamonds Botswana (Proprietary) Limited
- Steinmetz Diamonds Botswana (Proprietary) Limited
- Safdico Botswana (Proprietary) Limited
- Rand Diamonds Botswana (Proprietary) Limited (Tiffany Group)

## 2007

- Pluczenik Botswana (Proprietary) Limited
- MotiGanz Botswana (Proprietary) Limited
- Lazare Kaplan Botswana (Proprietary) Limited
- Leo Schachter Botswana (Proprietary) Limited
- Hearts & Arrows Cutting Works Botswana (Proprietary) Limited (Exelco Group)
- Eurostar Botswana (Proprietary) Limited
- Dia Holdings (Proprietary) Limited (belong to Star Diamonds Group)
- Shrenuj Botswana

Gem Diamonds acquired Gope Exploration from De Beers and Xstrata in May

## 2008

March 18th, DTCB building opened – the largest rough diamond sorting and valuation facility in the world

## 2008

April 1st, Seretse Khama Ian Khama sworn in as fourth and current President of Botswana.

19th April, Dr. Gavin Lamont, Discoverer of the Orapa and Jwaneng Kimberlite Fields in Botswana, passes away aged 87

African Diamonds given mining licence to mine A/K6 on 10th October

DiamonEX Botswana opens Lerala Diamond Mine, but then bought out by UK based Mantle Diamonds Ltd

Botswana Diamond Technology Park opened, includes diamond polishing, cutting and Jewellery companies as well as Gemological Institute of America

Botswana Diamond Hub Established

# Introduction

## Botswana **Diamond Timeline**

**2009**

First DTCB "sight" held on 23rd June

12th June, inaugural DTCB Shining Light Diamond Jewellery Design Awards

**2010**

B/K11 Mine commences Production

November, Shrenuj Diamond cutting and polishing factory opens in Broadhurst

**2010**

November, B/K11 commences production

A/K6 (Karowe Mine) commences construction

10th December, President Sereste Khama, Ian Khama launches the Jwaneng Mine Cut 8 project

**2011**

First non-Debswana rough diamonds sold in Botswana (B/K11-Firestone)

Lerala Mine reopens, owned by UK based Mantle Diamonds Ltd

Shrenuj Diamond Jewellery factory opens at Diamond Technology Park in March.

May, mining licence granted to Gem Diamonds for Gope Mine – name changed to Ghaghoo Mine. Production scheduled to commence in 2013

**2011**

16th September, Government signs new 10 year sales agreement with De Beers, backdated to 1st January, De Beers agreeing to migrate Diamond Trading Company International (DTCI) from London to Gaborone by the end of 2013. Government also now allowed to independently sell 10%, growing to 15%, of Debswana production

After over 80 years in the diamond business, on 4th November, Oppenheimer family announce sale of their 40% share in De Beers to Anglo American and the Botswana Government given the opportunity of increasing their stake in De Beers from 15% to 25%

## 2011

November, Opening of Laurelton diamond factory – the third largest diamond cutting and polishing factory in Botswana

December – Announced appointment of Tabake Kopedi as first Motswana Managing Director of DTCB and increase in number of Sightholders (suppliers of choice) from 16 to 21, commencing March 2012

## 2012

March – Formation of De Beers Aggregation Botswana and Okavango Diamond Trading Company

May 3 – A/K6 (Karowe Mine) officially opened

*Night time loading and hauling (Debswana)*

# Introduction

*Sunset at Orapa Mine (Debswana)*

The eleven chapters of this book, the first of its kind, dedicated wholly to the story behind Botswana's diamond success, follows the diamond from "Mine to Finger", or "Rough to Store", in what is known in the diamond business as the "Diamond Pipeline". The diamond pipeline includes access and exploration, growth, mining, ore extraction process, sorting, selling rough, cutting and polishing, jewellery manufacture, marketing diamond jewellery and diamond retail (Debswana, 2010). However, the phrase "Pipeline" had a different meaning when it was coined by the former De Beers Central Selling Organisation (CSO) in the 1980's. Wannernburgh and Johnson (1990) write "Performing its traditional stabilising role, the CSO maintains a fine balance between supply and demand, regulating the flow of diamonds through what is known in the trade as the "pipeline", which runs from Charterhouse Street (the HQ in London) successively to dealers and manufacturers, jewellery makers, wholesalers and retailers, and ultimately the consumers of diamond jewellery".

I describe what I term the seven periods or phases in the history of Botswana's diamonds:-

1) **1954 - 1966. "The lean and mean period"** Includes De Beers (Kimberltic Searches) preparation for establishing prospecting units in Bechuanaland, establishing its base in Lobatse and slowly building up resources and capability. This was a lean and mean period for diamonds; the only two kimberlites, discovered near Mochudi, turned out to be non-diamondiferous.

2) **1967 - 1974.** I term this **"the wonderful discovery era"** and this period is in stark contrast to the previous one. Over 50 Kimberlites, mostly diamondiferous, were discovered by De Beers Prospecting, during this five year period, including Orapa A/K1, the second largest mined kimberlite in the world (after the Madui pipe at the Williamson Mine in Tanzania) and also 2125 D/K1 and D/K2, later to become the Debswana Letlhakane Mine and 2424D/K2 at Jwaneng, later to become Jwaneng Mine, the richest diamond mine in the world. There was obviously no turning back.

3) **1971 - 1982.** This was the **"decade of big mine commissionings and openings"** in Botswana which firmly placed Botswana on the world diamond scene, later to become the largest diamond producer by both volume and value.

4) **1980 - 2011.** I term this period **"Prospects re-visited".** Over the last 30 years, original De Beers kimberlite discoveries in the 1960s and 1970s, which were abandoned at the time for being uneconomic, have been re-visited by others, using more advanced evaluation technology now available, and have shown the kimberlties to be now economic. Such examples are kimberlite pipes B/K 11 and A/K6 in the Orapa Kimberlite field, discovered by De Beers in 1967 and 1970 respectively, and which have now been revisited by Firestone Diamonds in 2008 and Boteti Mining in 2009, and both are now diamond mines.

5) **2003 - 2011** – **"The smaller mines period".** Damtshaa Mine was opened in 2003. Lerala Mine in 2008, B/K11 Mine in 2010, A/K6 in 2012 and Ghaghoo Mine is scheduled to open in 2013.

6) **2006 - 2011** – This is termed the **"The diamond valuation and trading consolidation period".** It saw the establishment of the Diamond Trading Company Botswana (DTCB) in 2006, which replaced the former Botswana Diamond Valuing Company (BDVC), the selection of Botswana's first diamond "Sightholders" (16) in 2007 and the construction of the DTCB HQ in 2008, the largest rough diamond sorting and valuation facility in the world. In 2011, for the first time, it was also agreed that the Botswana Government would independently sell 10% of the Debswana run-of-mine production in order to verify diamond markets. De Beers also agreed to re-locate Diamond Trading Company International (DTCI) from London to Gaborone by the end of 2013.

7) **2008 - Present** – I have termed this same period **"Towards Greater Diamond Beneficiation".** The Ministry of Minerals, Energy and Water Resources has stated that *"Our aim and desire for diamond mining is to see all activities of the diamond pipeline undertaken sustainably in Botswana"*. As a result, the Diamond Technology Park was opened in 2008 along with the Botswana Government's Diamond Hub and Diamond Center. In 2011, Botswana became a full member of the International Diamond Manufacturing Association and hosted its annual conference in Gaborone.

*DTCB Diamond Sorting and Valuation HQ, Gaborone (DTCB, 2008)*

In the next chapter, we take a look at how revenue from diamonds has transformed the Republic of Botswana from one of the poorest countries in the world to a modern and rapidly developing nation.

HOTEL
DIXONS Hotel Bottle Store

# 2 Botswana Before and after Diamonds

Botswana is a land locked country surrounded by South Africa to the south, Zimbabwe to the northeast, Zambia to the north and Namibia to the west.

It occupies an area of 581,730 km$^2$, and according to the 2011 census, has a population of just over two million, with a very low density of 3½ people per square kilometer. It is larger than France and about the same size as the state of Texas in the United States of America. English is the official language of Botswana although 78% of the population speak Setswana, the national language.

*Dixons Hotel and Shopping Mall at Mafeking, former capital of Bechuanaland until 1966, when Gaborone became the capital of Botswana. (BOPA)*

# Botswana Before and after Diamonds

*Old Gaborone Hospital, 1893 (Botswana National Archives and Records)*

*Bokamoso Private Hospital (M. Brook, 2011)*

At the time of Independence in 1966, the population stood at about 550,000 and there was only about 6km of tarred road, 3km in Lobatse, constructed for a visit to Botswana of the Queen of England's mother, and 3km in Gaborone at the international airport. Road access in the sandy south of the country was especially difficult, and without four-wheel drive vehicles, the most reliable and sturdy form of transport for police patrols were camels, acquired from the then German occupied South - West Africa (now Namibia). Today, after several generations, there are over 200 camels in the Tshabong area which have been placed in a Camel Eco-Tourism Park, a partnership between the Botswana Tourism Organisation and the local Tsamama Community Trust.

The nation was originally inhabited by the San people (also known as Bushmen or Basarwa), but they constitute only a very small portion of the population today. Beginning in the 1820s, the region was disrupted by the expansion, from South Africa, of the Zulu people and their offshoot, the Ndebele. However, Khama II, chief of the Ngwato, curbed the depredations of the Ndebele and established a reasonably unified Gammangwato.

A new threat arose in the late mid 19th century with the incursion of the Boer farmer regiments from neighbouring Transvaal in South Africa. After gold was discovered in northern parts of Botswana in 1867, the Transvaal government sought to annex parts of Botswana. Although the British forbade annexation, the Boers continued to encroach on what was known as tribal native lands during the 1870s and 1880s.

*Pre-diamonds school class room*

*Shakawe Senior Secondary School (BOPA, 2011)*

German colonial expansion in South West Africa (Namibia) caused the British to re-examine their policies, and, at the urgent request of the three Chiefs Khama, Bathoen and Sebele, they established (1884–85) a protectorate called Bechuanaland. The southern part of the area was incorporated into the Cape Colony in 1895.

Until 1961, Bechuanaland was administered by a resident commissioner at Mafikeng, in South Africa, who was responsible to the British High Commissioner for South Africa, based in Pretoria. Relatively little development took place during the Protectorate period, and it was intended that Bechuanaland would eventually be incorporated into the Union of South Africa.

The rise of the National Party in South Africa in 1948 and its pursuit of apartheid turned British opinion against the incorporation of Bechuanaland into South Africa, and South Africa's attempts at annexation were strongly countered by British insistence that Bechuanaland's inhabitants be consulted first.

Eventually it was acknowledged that Bechuanaland would become an independent nation, and there was some limited investment in social and economic infrastructure from the mid-1950s onwards. Britain granted Bechuanaland internal self-government in 1965 and full independence as Botswana on September 30, 1966, when the first President, Seretse Khama was inaugurated. He was later to be knighted by Queen Elizabeth II and became known as Sir Seretse Khama. His eldest son, Ian Khama, is the current President of Botswana, only the fourth President to take office in a period of 45 years.

# Botswana Before and after Diamonds

*Botswana Police Services Camel Corps, Tshabong (BOPA, 1966)*

*First Female Motswana Police Helicopter Pilot of Eurocopter AS350 (BOPA, 2009)*

Botswana was one of the poorest countries in the world in 1966 with a per capita income of about US$80. The poverty situation was accentuated by several years of drought, which coincided, with the achievement of independence. One third of the national cattle herd died, while one fifth of the population received famine relief from the international community.

For the first five years of its political independence, Botswana remained financially dependent on Britain to cover the full cost of administration and development. The planning and execution of economic development took off in 1967-71 after the discovery of diamonds at Orapa.

*The first De Beers aeroplane to land at Orapa, 16th October (M. Marx, 1967)*

*Debswana aircraft today (M. Brook, 2011)*

To begin with there were few schools and medical facilities, mostly restricted to the new capital town of Gaborone, but soon they existed all over the country. Immediately after independence, the main export and virtually only revenue driver for government was beef.

However, following the discovery of diamonds, Botswana recorded the highest average economic growth rate in the world averaging 8% per year from the 1980s to date. Growth in formal sector employment has averaged about 10% per annum over Botswana's first 30 years of independence. This sustained growth transformed Botswana from being one of the poorest countries in the world in 1966 to middle-income status by the 1990s. GDP per capita income in 2010 was US$15,488, the third highest in Africa (after the Equatorial Guinea, $18,143 and Seychelles, $23,428).

*Archibald Mogwe, Minister of Mineral Resources and Water Affairs between 1985 and 1994 (Botswana National Archives and Records)*

*"At the time of independence skeptics gave us little chance of economic survival"*, said Archibald Mogwe in 1996, Minister of Mineral Resources and Water Affairs between 1985 and 1994. *"The diamond mines were a godsend, and rapidly transformed the economy, increasing the revenue of the country from less than ten million Pula a year at independence to over a billion today".*

Indeed, at the time of Independence, the country was purportedly the third poorest country in the world!

However, according to Even - Zohar (2007), an author and renowned expert in the field of diamonds, *"Sound macroeconomic policies and the prudent use of diamond revenues have made Botswana one of the fastest growing economies in the world over the past 25 years"*. The revenues derived from Botswana's diamonds have provided a platform upon which to build a new nation from scratch, which is exactly what Botswana has done over the last 45 years.

Botswana has strived towards meeting the 2000 Millennium Development Goals of reducing poverty. Since independence in 1966 Botswana has adopted an approach of development planning and government which has since then, produced a series of National Development Plans (NDPs) starting with Botswana's Transitional Plan for Social and Economic Development prepared in 1965. National Development Plan 10, or NDP 10, covering the period 2009 to 2016, is the tenth in the series of NDPs. National Development Plans guide the overall development of the country.

**Diamond Exports 1998-2011**

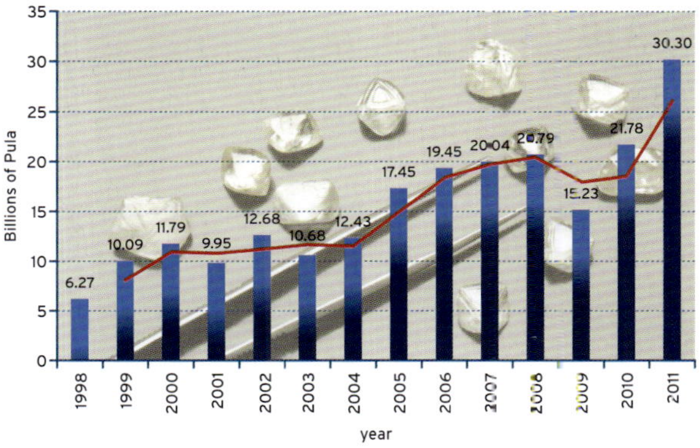


*Revenue from Diamond exports 1998 - 2011 (Data Source: Central Statistics Office)*

# Botswana Before and after Diamonds

*Gaborone City Center and Parliament Buildings, July 1980 at the time of Sir Seretse Khama's (first President of Botswana), lying in state. (BOPA, 1980)*

In 2006, former President of Botswana, Festus Mogae, stated that *"Between independence in 1966 and 2005, real gross domestic product grew by an average of more than 7 percent per annum. GDP per capita was over $4,000.00 U.S. dollars by 2005. The national budget has grown from under $3 million in 1966 to $4 billion in 2006. Exports have also grown to over $2.5 billion by 2005. Formal sector employment has risen close to 300,000 last year. Life expectancy at birth rose from 48 years in 1966 to 65.3 years in 1991 before declining to 55.6 years in 2001 partly as a result of the HIV/AIDS pandemic. Primary school enrollment rose from 66,100 in 1966 to 327,600 last year. Access to health, education, transport, telecommunications and other services have improved dramatically. The Botswana economy which is premised on free market principles has undergone a significant transformation since independence. Agriculture accounted for 40 percent of GDP in 1966. It now accounts for less than 3 percent. Mining at over 33 percent is now the larget contributor to gross domestic product. The Botswana economy is open with exports plus imports constituting over 80 percent of gross domestic product. At independence exports primarily consisted of beef but diamond exports soon took over and accounted for over 80 percent of total exports in 2005. Then of course, the multiply effects of mining on the economy is significantly much more than these figures suggest as mineral revenues are used to develop other sectors of the economy such as education and health. Botswana graduated from the list of 24 least developed countries in 1992".*

Botswana has a flourishing multiparty constitutional democracy. General elections are held every five years. Each of the elections since independence has been freely and fairly contested and has been held on schedule. The country's opposition parties participate freely in the political process. The openness of the country's political system has been a significant factor in Botswana's stability and economic growth.

*Government Enclave, Gaborone (Illustrative Options, 2007)*

In terms of cash flows from diamond revenues, the government receives a 10% royalty on Debswana's and other diamond mining companies gross revenue, a company tax of 10%, an additional income tax of 15% and, finally, the withholding tax of 15% on company dividends. It is therefore not surprising that revenues from diamond mining have consistently contributed to over 30% of Gross Domestic Product and over 60% of total government revenue. Government revenue from diamonds has increased from 6.27 billion Pula in 1998 to 30.3 billion Pula in 2011, an average annual increase of 6%.

Today, Botswana is widely considered to be one of the leading countries in Africa in respect of governance. This reflects the generally high quality of public institutions, an independent legal system, and a low level of corruption, all of which have been sustained over a long period of time. This achievement has been reflected in a number of international surveys and assessments like the World Economic Forum (WEF) and the Ibrahim Index of African Governance.

The WEF assesses the quality of institutions as the first pillar of its Global Competitiveness Index. Of the 19 indicators measured under the institution's pillar, Botswana scores in the top 50 countries globally on 16 of them. Botswana does particularly well in rankings for (lack of) wastefulness in government spending, public trust in politicians, diversion of public funds, favouritism in decisions of government officials, and efficiency of legal framework in challenging regulations.

In 2010, the Ibrahim Index of African Governance, which measures the quality of governance in sub-Saharan Africa's 48 states - on the basis of safety and security, rule of law, transparency and corruption, participation and human rights, sustainable economic opportunity, and human development - ranked Botswana as the third best, an improvement from the fourth best governed country in 2009. Botswana's overall score was 75.9, after Mauritius and Seychelles which scored 83, and 78.5 respectively. To this extent, Botswana is amongst the top performers on the African continent.

Botswana's impressive economic performance has been the basis for widespread social development. Government revenues, primarily from diamond mining, have been used to fund an extensive social welfare programme, particularly in the form of public education, health provision, water reticulation, roads and other infrastructure. Given that the level of social and physical development was extremely low at the time of independence, considerable achievements have been made in achieving virtual universal primary and junior secondary education, health care (88% of the population live within 8km of a health facility, and trained health personnel attend to 99% of births) and access to clean water supplies (97% of the population have access to safe drinking water). It is only the most remote areas that state provision of these basic services may not be available (Capital Resources, 2011).

# Botswana Before and after Diamonds

*First Fire Engine in Botswana being presented to the Mayor of Gaborone, Mayor Dambe (BOPA, 1968)*

Diamonds have provided the country with an opportunity to evolve from its pre-independence financial situation to present day economic stability. Botswana is rated by the Berlin-based non-profit organization, Transparency International, as the most transparent country on the African continent.

Despite a flat demand for diamonds in the early 1970s when Botswana truly began to develop, the growing global diamond market has made Gaborone, the capital, the fastest growing city on the planet, with the best credit rating per capita in Africa (World Report Inc., 2006).

Botswana strongly believes in Diamonds for good! Former President, Festus Mogae, stated *"For our people, every diamond purchase represents food on the table; better living conditions; better healthcare; potable and safe drinking water; more roads to connect our communities; and much, much more."*

Most companies now involved in the diamond business in Botswana have developed Corporate Social Investment programmes. For example, the first donation Debswana Diamond Company made was 25 South African Rands to Our Lady of the Desert primary school in Francistown over forty years ago. Today, the company reserves approximately US$ 2 million annually for social investment.

Botswana's exports are dominated by diamonds, which have accounted for around 65% of the total in recent years; other important exports are copper-nickel, gold, beef, soda ash and textiles. Diamonds historically have been sold through De Beers marketing channels in the United Kingdom, but revenues are received in US dollars. These marketing channels are progressively being transferred to Botswana, and a portion of diamonds is now sold to local cutting factories, which export cut and polished rather than rough diamonds.

*Gaborone Fire Engine (M. Brook, 2011)*

Other exports are spread across a range of international markets, primarily in Europe and Southern Africa. The share of diamonds has been declining in recent years as copper and nickel have risen and exports have become more diversified. The export of services, especially tourism, have also been increasing (Capital Resources, 2011). Botswana is a signatory to numerous regional and international trade agreements. Perhaps the most important of these is the Southern African Customs Union (SACU), under which Botswana enjoys free trade with South Africa, the largest and most important regional market.

Tourism is an increasingly important industry in Botswana, accounting for approximately 11% of GDP, and has grown at an annual rate of 14% in the past eight years. One of the world's unique ecosystems, the Okavango Delta, is located in Botswana and is the largest inland freshwater delta system. The country offers excellent game viewing and birding both in the Delta and in the Chobe Game Reserve–home to one of the largest herds of free-ranging elephants in the world. Botswana's Central Kalahari Game Reserve also offers good game viewing and some of the most remote and unspoiled wilderness in southern Africa.

More than half of Botswana's population lives in rural areas and is largely dependent on subsistence crop and livestock farming. Agriculture meets only a small portion of food needs and makes up only 2.3% of GDP–primarily through beef exports–but it remains a social and cultural touchstone. Cattle raising in particular dominated Botswana's social and economic life before independence. The national herd is estimated at between two and three million, so there are a lot more cattle than people!

Botswana is now a modern, independent African State which has both political and economic stability. The revenues gained from the mining sector (especially diamonds) have been used prudently by successive governments to establish the developmental platform required to advance the nation towards prosperity, by reducing poverty. The benefits of diamonds have been spread across the country, despite its vastness and sometimes difficult access. A modern international airport in Gaborone, and other international airports at Francistown, Kasane and Maun have enabled Botswana to strengthen its regional and international communications which have also allowed a thriving tourism sector to develop. By 2001, the total length of roads was about 20,000km (Botswana Dept. Surveys & Mapping, 2001), and increased to 36,500km by 2010.

Botswana is fortunate to have been able to develop rapidly from the benefit of diamond revenues.

In the next chapter, Botswana's favourable geology is shown to be the reason for prolific kimberlite occurrences and the discovery of diamonds in the country.

# 3 Geological Setting and Formation of Diamonds

Geologically, Botswana remains relatively unexplored and undocumented, largely because of the fact that over two thirds of the country is blanketed by thick sand deposits of the Kalahari Desert.

The lack of geological outcrop has made mapping difficult. In recent years however, technological advances in geophysical exploration have meant that hitherto hidden geological formations are attracting renewed attention from mining companies. Over the past decade, there has been a surge of exploration activity, not just for diamonds, but also for base metals (copper, nickel, gold, uranium etc) and coal, of which Botswana has centuries of supplies from currently un-mined deposits.

*DTC Gems (DTC, 2007)*

# Geological Setting and Formation of Diamonds

### BOTSWANA'S KIMBERLITE FIELDS

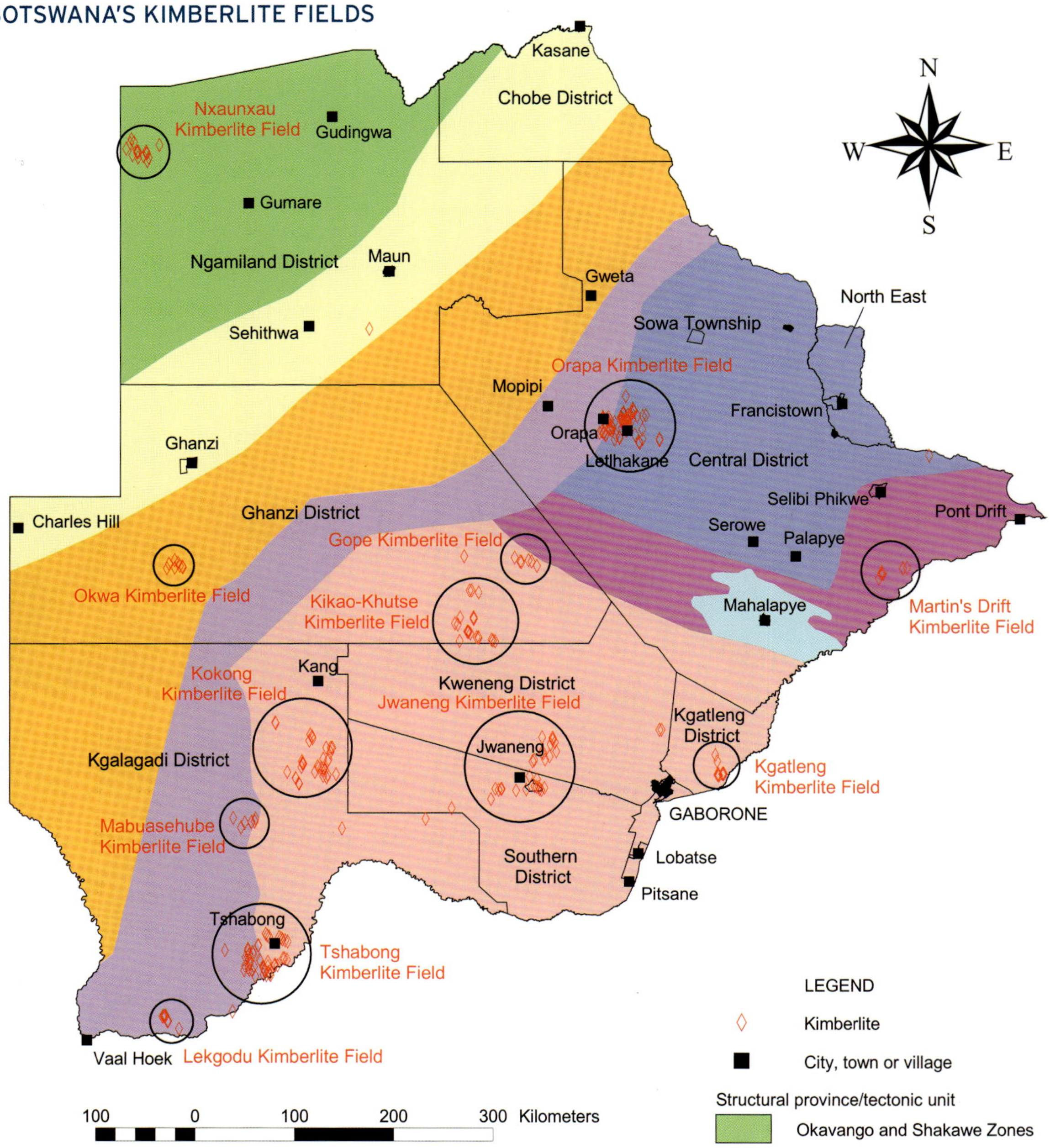


| KIMBERLITE FIELD | NUMBER | KIMBERLITE FIELD | NUMBER |
|---|---|---|---|
| Nxaunxau | 27 | Jwaneng | 28 |
| Okwa | 8 | Mabuasehube | 9 |
| Orapa | 81 | Tshabong | 86 |
| Gope | 9 | Lekgodu | 8 |
| Kikao-Khutse | 23 | Kgatleng | 11 |
| Kokong | 76 | Other | 9 |
| Martin's Drift | 8 | **TOTAL** | **383** |

The geology of Botswana can be categorized into three main underlying groups of rock unit and sequence of geologic events:

- **Phanerozoic.** This comprises rocks which are between 500 million years old and those of the present day. Minerals that have economic potential and that are found in this time horizon, include coal, hydrocarbons, limestone, *diamonds* and brickearth.

- **Proterozoic.** Comprises rock formations that are between 500 million years old and 2.5 billion years in age. The rocks that occur in this time horizon could potentially be a source of economic minerals such as platinum, copper, silver, lead, zinc, asbestos, mineral pigments and limestone.

- **Archaean.** This is the oldest of the three categories and comprises all rock formations older than 2.5 billion years. Archaean rocks can be host to economic minerals such as gold, silver, copper, nickel, limestone and silica.

Thus the geological framework of Botswana includes cratonic regions comprising parts of ancient stable blocks of Archean metamorphosed rocks, mobile belt regions which have suffered tectonism at various periods during the Proterozoic, and platform regions on which un-metamorphosed Proterozoic and Phanerozoic sedimentary and volcanic rocks have been preserved.

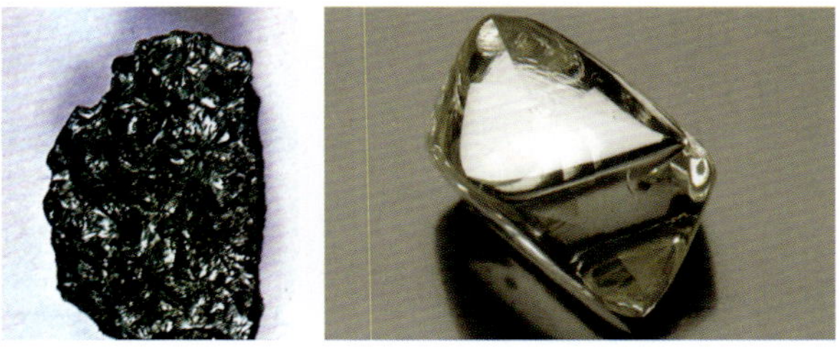

*Diamond and Graphite have the same carbon chemical composition but Graphites hexagonal crystal system compared to diamond's cubic system is the reason for the complete contrast in their physical and structural properties*

The Archaean 'basement' sequence only outcrops in the eastern part of Botswana. Both the Kaapvaal and Zimbabwe Cratons were formed during different stages with the Kaapvaal Craton being the older of the two. The 'intercratonic' Limpopo belt separates the two. The cratons are composed mainly of granitoid and supracrustal lithologies with associated gneisses and the greenstone belts present in the intercratonic domain.

These predominantly metamorphic lithologies have other Archaean supracrustal sequences and igneous intrusions, which are relatively underformed, associated with them. Examples of this are the Lobatse Group volcanics and the Kanye Formations. The youngest of these is the Mahalapye complex, which is an assemblage of gneisses, migmatites and variably foliated plutonic rocks within the Limpopo belt. During the late stages of the Archaean, the Kaapvaal, Zimbabwe and Limpopo units acted as a single, stable crustal block (collectively referred to as the Kaapvaal-Zimbabwe Craton).

The Proterozoic sequence succeeds the Archaean rocks of the Kaapvaal-Zimbabwe Craton. The oldest is the Transvaal Supergroup followed by the Otse (Botswana) and Waterberg Supergroups and various doleritic dykes with other intrusions. These sequences were probably deposited about 1.2 to 2.5 billion years ago.

Proterozoic mobile belts are distinguishable along the western margins of the Kaapvaal- Zimbabwe craton. The oldest of these are the Kheis and Magondi orogenic belts which are 1. 8 to 2 billion years old. Archaean-age lithologies can be found in parts of this mobile belt.

The major phases of Proterozoic mobile belt formation occurred deep within the cratonic interiors. This included extensive folding and faulting of the sequences and intrusion by basic igneous bodies of various compositional types.

# Geological Setting and Formation of Diamonds

**REGIONAL LOCATION OF DIAMOND MINES (Field, M *et al*, 2008)**

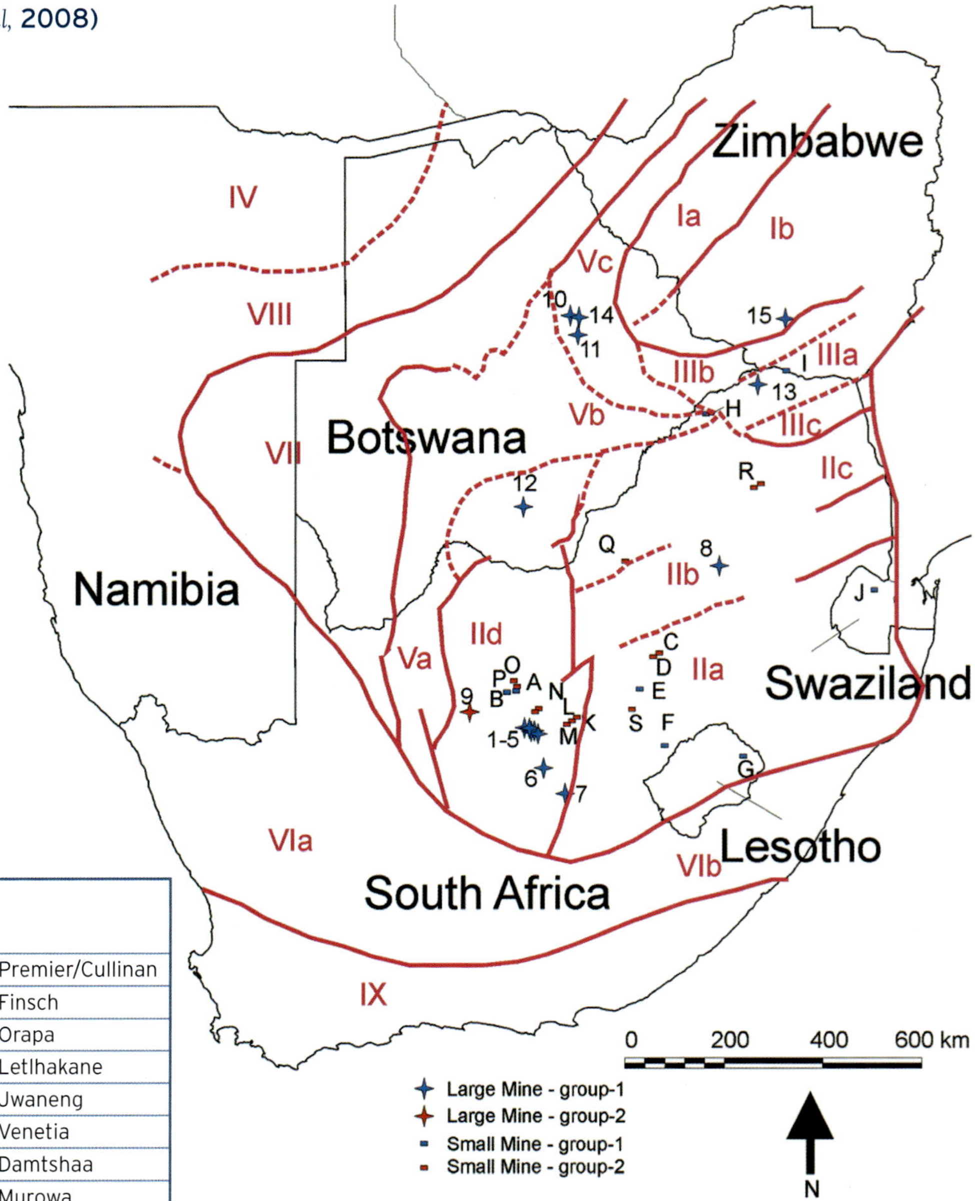


**LARGE MINES**

| | | | |
|---|---|---|---|
| 1 | Kimberley | 8 | Premier/Cullinan |
| 2 | De Beers | 9 | Finsch |
| 3 | Bultfontein | 10 | Orapa |
| 4 | Dutoitspan | 11 | Letlhakane |
| 5 | Wesselton | 12 | Jwaneng |
| 6 | Koffiefontein | 13 | Venetia |
| 7 | Jagersfontein | 14 | Damtshaa |
| | | 15 | Murowa |

**SMALL MINES**

| | | | | | | | |
|---|---|---|---|---|---|---|---|
| A | Frank Smith | G | Letseng | M | Blaauwbosch | S | Star |
| B | Newlands | H | The Oaks | N | Loxtondal | | |
| C | Voorspoed | I | River Ranch | O | Bellsbank | | |
| D | Crown/Lace | J | Dokolwayo | P | Doornkloof-Sover | | |
| E | Kaalvallei | K | Roberts Victor | Q | Swartruggens | | |
| F | Monastery | L | New Elands | R | Marsfontein-Klipspringer | | |

One of these, the Molopo Farms Complex, is similar in age to the Bushveld Igneous Complex in South Africa having been emplaced about 2.05 billion years ago. Late Proterozoic sediments in western Botswana, filled various basins that were established during the final stages of the Proterozoic tectonothermal cycle, in what was called the Damara Orogenic Belt. The Congo Craton with overthrust late Proterozoic rocks, is the main assemblage in the northwestern part of the country.

Finally, the Phanerozoic successions, covering about 75% of Botswana's surface area, are those of the Karoo Supergroup and the Kalahari beds. The Kalahari Group consists of loose, unconsolidated aeolian sand that is generally more than 25 metres deep, but in the central Kalahari, can reach thicknesses of greater than 80 meters. Indurate sand with calcareous or siliceous cement underlies the soft sands, and, at the base of the aeolian sand, there is often a calcrete - silcrete duricrust horizon. With increasing depth, the sandstone becomes redder and has increasing clay content. The base of the unit is often defined by the presence of a red marl or clayey sandstone, but infrequent basal gravels do also exist in certain areas.

All of Botswana's diamond mines recover kimberlite, the host rock for diamonds, and are located on or close to the Archaen Cratons and according to Field et al. (2008), *"it is apparent that the craton and its associated sub-continental lithosphere played an important role in providing the right environment for diamond growth and for the formation of the kimberlite magmas that were to transport them to surface. However it is often not easy to determine whether a particular site is "on craton"*.

The reason for the association between Archaean basement and diamondiferous kimberlites has been explained theoretically by consideration of the structure of the cratons and the temperature/ pressure relationship between graphite and diamond. Diamonds form and are preserved in a high pressure environment present in nature at depths of over 120 kilometers beneath the earth's surface. In most parts of the Earth, the temperatures at this depth are too high for diamonds to form. However Archaean cratons have relatively cool lithospheric roots in which there exists a downward deflection of isotherms and a corresponding upward deflection of the diamond stability field (De Beers Diamond Geology).

## SIGNIFICANCE OF CRATONS

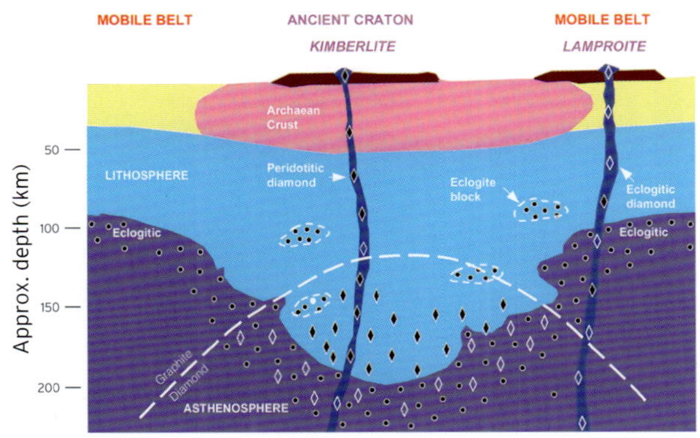


*Significance of Cratons for Kimberlite Emplacement (De Beers Diamond Geology)*

This region of high pressure and relatively low temperature (less than about 1200°C) provides a "window" in which diamonds can form and be preserved. Kimberlitic magmas are generated at or below these depths (as evidenced by their xenoliths), and may "sample" the lithospheric roots, thus collecting diamonds en route to surface. Kimberlites formed away from the craton do not sample the diamond window, and thus are unlikely to be diamondiferous.

# Geological Setting and Formation of Diamonds

*Hypabyssal kimberlite texture (Hamilton, 1975)*

*Kimberlite Breccia at Orapa Mine (M. Brook, 2009)*

*Debswana Damtshaa Mine Kimberlite Boulder (M. Brook, 2010)*

According to Hamilton et. al. (1975), kimberlite is a bluish, greenish or black igneous rock with a range of minerals, including olivine, ilmenite, phlogopite, chrome diopside, and garnet which tend to be rounded and perhaps broken, forming phenocrysts. Its structure often contains xenoliths. Calcite may also be abundant. Morphologically, they attain the shape of a carrot, tapering downwards.

From work done on kimberlites in many parts of the world, three broadly distinct vertical zones are recognised, the crater, diatreme, and root zones. The crater represents the uppermost portion of the pipe and is characterised by well-bedded, poorly consolidated sediments with chaotic debris-flow deposits and pyroclastics. These sediments were sometimes laid down in a lake in the center of a volcanic crater e.g. Orapa Mine.

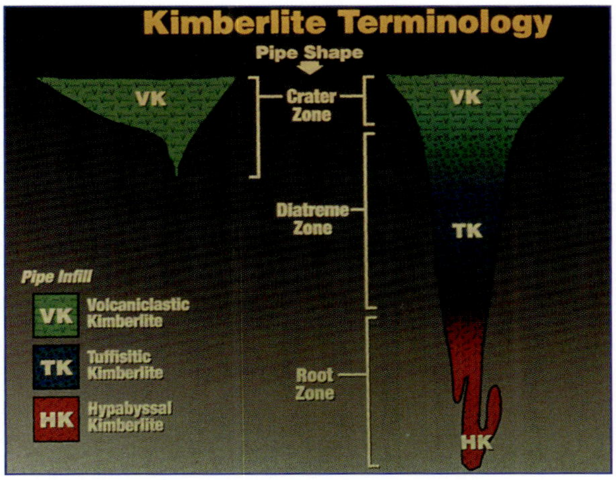


*Kimberlite Terminology (Field & Scott Smith 1998)*

*Different Orapa A/K1 Kimberlite polished sections from the volcaniclastic, tuffisitic and hypabyssal facies (M. Brook, 2011)*

*South pipe/quartzitic shale country rock contact at Jwaneng Mine (M. Brook, 2009)*

The middle part, the diatreme, is volumetrically the most significant, and comprises an easily weathered breccia consisting of angular country rock xenoliths and fragments of mantle-derived material set in a fine-grained matrix. This is known as a tuffisitic kimberlite breccia or TKB. The first two parts of the kimberlite profile were termed "yellow ground" by Kimberly, South Africa mine workers in the 19th Century.

The final part, the root zone, or "blue ground", is composed of magmatic or hypabyssal material usually porphyritic in appearance, containing macrocrysts of olivine and phlogopite set in a fine-grained matrix, often with xenocrystic garnet, ilmenite, spinel and chrome-diopside. Mining difficulties may be experienced in the hypabyssal facies due to the irregular shapes of the intrusion.

The presence and degree of preservation of these three different zones depends upon the level of erosion, the volatile content of the erupting magma and the stability and nature of the country rock. The large, economically important kimberlites at Orapa and Jwaneng have suffered very little erosion, and their crater facies were preserved. Other pipes in South Africa at the Kimberley, Jagersfontein, and Koffiefontein mines, are eroded down to the diatreme zone.

Volcanic eruptions pass through the diamond bearing rocks deep under the earth at depths of between 100 and more than 200 kilometers and sometimes transport the diamonds (but always only as very rare included elements) to the surface where the rock hardens to form the kimberlite. Often clusters of up to 50 pipes are found. In both the Orapa and Tshabong clusters or fields, over 80 kimberlites have been discovered to date. Kimberlites most commonly occur as carrot-shaped pipes, as dykes, or more rarely as sills.

# Geological Setting and Formation of Diamonds

*Diamond in drill core - North pipe – Jwaneng Mine (M. Roberts, 2012)*

The size of Botswana kimberlites varies, with the largest at 146 hectares, being the M1 kimberlite pipe in the Tshabong Field, in the Kgalagadi District, Southern Botswana.

Diamond grades within a kimberlite can vary from zero, in a non diamondiferous pipe, to between 3 carats per hundred tonnes and greater than 200 carats per hundred tonnes in an economic diamondiferous pipe. Other factors which are important in assessing the economics of diamond pipe mining are; the average value per carat and the tonneage available.

Diamond resources can be classified as primary (kimberlites and lamproites) and secondary (alluvial and marine). The dispersal of diamonds from their primary, in-situ source takes place into streams and rivers and ultimately to the coast and is generally accompanied by an increase in average value per carat, as flawed stones are progressively destroyed with greater and greater transport. Diamonds are known to occur in a variety of rocks, including high-pressure metamorphic rocks such as the garnet-biotite gneisses of northern Kazakhstan, alpine type peridotites, and meteorites.

Kimberlites with preserved crater facies rocks are much rarer, but in a few cases, such as Orapa during early mining days, are important high grade and high tonnage deposits.

*Jwaneng South Pipe, (M. Brook, 2012)*

*Orapa A/K1 Brecciated Kimberlite polished section (M. Brook, 2011)*

However, the only known economically significant primary sources of diamonds are kimberlite and lamproite igneous rocks. No diamondiferous lamproites are known in Botswana where the primary sources mined are kimberlite pipes and dykes.

In kimberlite-hosted deposits, diamonds occur mainly as sparsely dispersed, mantle-derived xenocrysts and diamondiferous mantle xenoliths in the kimberlite matrix (Kjarsgaard, 1996). Economic quantities of diamonds are mainly found in kimberlite diatremes.

*Orapa Mine Basalt Breccia Kimberlite polished section (M. Brook, 2011)*

# Geological Setting and Formation of Diamonds

*Orapa A/K1 Tuffisitic Kimberlite rock polished section with calcite veins (M. Brook, 2011)*

Kimberlite is defined as a volatile-rich, potassic, ultrabasic igneous rock that occurs as small and large volcanic pipes, dykes and sills. It has an inequigranular texture resulting from the presence of macrocrysts (phenocrysts and xenocrysts) set in a fine-grained matrix. The mineralogy comprises olivine with several of the following: phlogopite, calcite, serpentine, diopside, monticellite, apatite, perovskite, and ilmenite.

Of all kimberltes found, less than 1% are of economic grade. Grade is specified as carats per hundred tonnes (cpht) kimberlite ore. In Botswana, the eight diamond mines grades range from 8.5 cpht at B/K11 in the Orapa kimberlite field to a massive 140cpht at Jwaneng Mine, which is by far the richest diamond mine in the world with mostly gem quality diamonds.

Two distinct types of kimberlite are recognised: Group I, or olivine-rich, calcite kimberlites and Group II, or micaceous kimberlites. These distinctive groups are derived from sources in the earth's mantle that are either slightly depleted (Group I), or enriched (Group II) with respect to light rare earth elements.

Only a small proportion of kimberlites carry diamonds; for example, of the 383 kimberlites and lamproites discovered in Botswana, only 75 (19%) are diamondiferous, but only 14 (3%) of the pipes are currently mined. Kimberlites occur in twelve different kimberlite fields in Botswana; the smallest field, the Kgatleng kimberlite field, contains eleven kimberlites (mostly of lamproite variety) and the largest field at Tshabong contains 85 kimberlites, followed by the Orapa field with 81.

*Pink tuffisitic kimberlite from Orapa Mine A/K1 (M. Brook, 2011)*

Across the boarder, in South Africa, of the approximately one thousand individual kimberlite intrusions known, only about fifty carry significant quantities of diamonds. Of these, many are considered sub-economic either because the quantity or quality of the diamonds, or the quantity of ore is insufficient (De Beers, Diamond Geology).

It is difficult to predict whether or not a kimberlite will carry diamonds without actually testing it. This is done by bulk sampling, a stage in between the kimberlite's discovery and commissioning a mine. The presence and quality of diamonds in a kimberlite can only be determined with confidence by the collection and processing of a large and representative sample.

*Orapa A/K1 Hypabyssal Kimberlite rock polished section with ilmenites and olivines (M. Brook, 2011)*

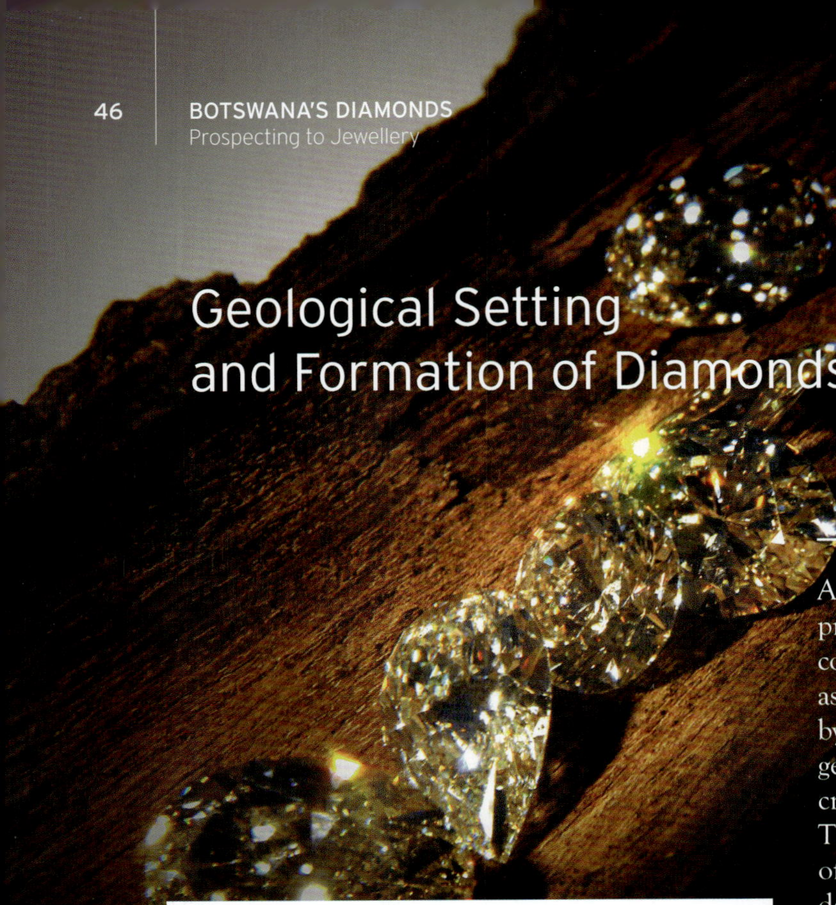

# Geological Setting and Formation of Diamonds

A diamond is distinguished from all other precious stones no less by its chemical composition, since it has the same mineralogy as graphite (pencil lead) and charcoal, than by its unique physical characters, for no other gem consists of a single element. It is a pure crystallized carbon with a cubic crystal system. The extraordinary difference in the appearance of diamond and that of other forms of carbon depends solely on the crystallization of the material and the physical characters consequent on this. Graphite has a hexagonal crystal system.

## DESCRIPTION OF A DIAMOND

A diamond is not the most valuable of all precious stones, yet it unquestionably exceeds all others in interest, importance and general noteworthiness. The name diamond originates from the ancient Greek αδάμας - adámas which means "unbreakable". In hardness, in the perfection of its clearness and transparency, in its unique constants of optical refraction and dispersion, and finally in the marvelous perfection of its luster, the diamond surpasses all other minerals on earth.

Diamonds commonly occur as octahedral crystals frequently of flattened habit, more rarely as cubes, often with curved faces. A diamond has a specific gravity of $3.5 g/cm^3$ and a hardness of 10 (maximum) on the Moh's scale – the hardest of all natural materials known to mankind, only a diamond can cut diamond. Its cleavage is octahedral and perfect and has a conchoidal fracture pattern. The mineral is transparent to opaque and may be colorless, yellowish, brown, red, blue, green and black. Gem quality diamonds are clear and grey to black and small opaque granular diamond is called "bort". Diamonds have a white streak and an adamantine luster, although uncut stones can appear greasy. A diamond is also the most effective heat conducting material, which also expands very little when subjected to high temperatures, unlike most other conducting materials.

It is also resistant to most acids and alkalis and is transparent to X-rays – a property which is fully utilized in the diamond recovery process. Diamonds actually luminesce or emit light when exposed to X-ray radiation.

A diamond is also "un-wettable", allowing stones to be trapped on a grease belt, which is still used in some operations as a means of diamond processing or recovery.

The weight of a diamond is measured in carats. The word carat derives from the Greek word 'keration' meaning fruit of the carob. The Carob tree (Ceratonia siliqua) is an evergreen tree, with an edible pod containing seeds, which is native to the Mediterranean region. The seeds of the Carob were used on precision scales as units of weight for small quantities of precious gemstones because of their uniform size. The weight of an average carob seed is 200 milligrams. The weight of one carat is precisely 200 milligrams, or 0.2 grams. In the Far East instead of carob beans, jewellers used rice grains to determine the weight of gemstones. Because of this fact, you may occasionally hear some jewellers referring to a 1-carat diamond as a "four grainer," which means in the past 1 carat was equal to four grains of rice. 1.5ct. would be a 6 grainer, and a 0.5ct. stone a 2 grainer.

The weight of smaller diamonds is often expressed as points, not carats. One carat is equal to 100 points. Thus, for example, a 10-point diamond has the weight of 0.1 carats.

Diamonds weighing less than 1/5ct. (less than 20 points) are referred to as Melee and stones this size are set using various techniques in jewellery. The diamond is one of the most perfectly crystallised of minerals.

Almost every single stone is bounded by regularly developed faces. The faces of diamond crystals differ from those of most other crystallized minerals, in that they are, as a rule, much curved and rounded instead of being perfectly plane, as is usually the case. Crystals of diamond belong to the cubic system. There are three basic forms of diamonds namely Octahedrons, Cubes and Dodecahedrons. Diamonds mostly crystallize either as octahedrons or cubes and this depends on the conditions at which the diamond is formed.

Primarily, diamonds form in an octahedral pattern, (hence the primary growth pattern). Then due to a change in conditions where they are formed, this results in a change in the growth pattern. Therefore, cubic growth is the secondary growth pattern of diamonds. Because of this factor, cubes have a tendency of inverted or concaved faces and a thick surface layer which makes it very difficult to see inside the diamond.

Botswana's Jwaneng Mine, the world's richest mine, is also commonly known for producing cubes.

Dodecahedron, also known as rhombic dodecahedron, is another form of diamonds which result from resorption or resolution.

Diamonds, as crystals, have symmetrical faces and fall under the isometric/cubic system. They have ribs - where two faces meet, hard corners - where four ribs meet and soft corners - where three ribs meet.

Giving recognition to the fact that Botswana is one of the world's leading producers of diamonds, it has been decided that the 11th International Kimberlite Conference of 2016 will be held in Botswana (Gaborone) for the first time.

Having shown how kimberlites occur within "Fields" within the stable craton geology, let's now explore the different prospecting methodologies which various companies have employed to discover so many kimberlites in Botswana.

# 4 Prospecting and Evaluation

With more than two thirds of Botswana covered by often-thick sands and other sediments of the great Kalahari Desert, De Beers Consolidated Mines (DBCM) in Kimberly, South Africa, always knew that diamond prospecting would have its challenges here.

Unperturbed, DBCM applied for their first diamond-prospecting licence in 1932. But because the Bechuanaland Mining and Prospecting Charter was still under preparation, it was a few years later that they were granted their first Crown Prospecting Grant. From Independence onwards, Botswana replaced the Crown grant with the State Grant which was issued for prospecting within specific prospecting licences.

*Diamond prospecting soil sampling team, Orapa (De Beers, 1967*

# Prospecting and Evaluation

## LOCATION OF DIAMOND MINES AND DIAMOND PROSPECTING LICENCES IN 2011

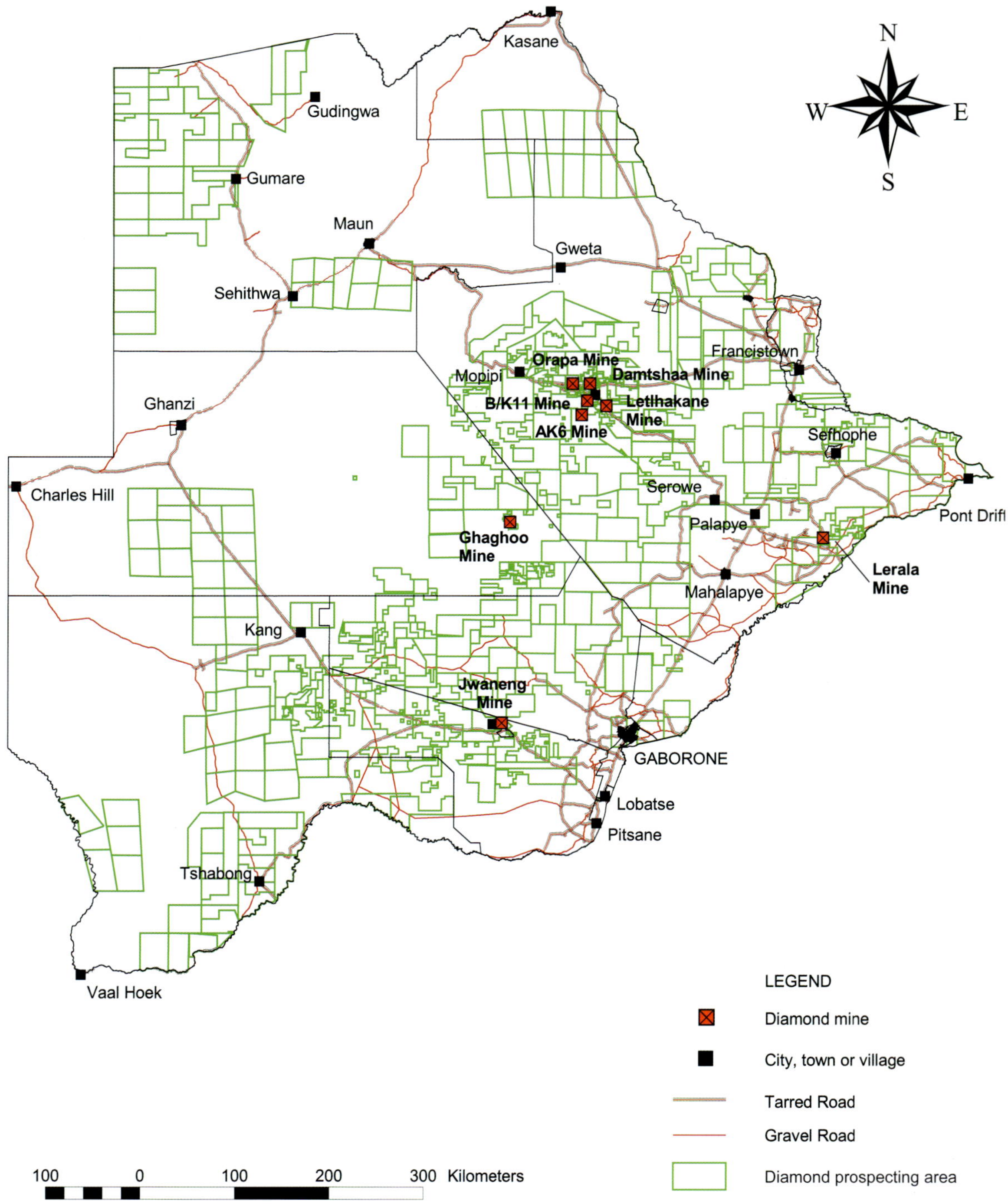

The conditions of a Crown Grant, as detailed in Part IV of the Draft Mining Act, 1932, were quite onerous. It stated that "The Crown grant may contain such limitations, terms and conditions as the High Commissioner may deem necessary, provided that the holder of a Crown grant shall:-

(a) Pay royalties to the Government of the Bechuanaland Protectorate in accordance with the provisions of Section 28; and

(b) Be entitled to peg such number of claims as may be prescribed in the Crown grant

**The Act also went on to state that:-**

(c) Crown grants may be made in respect of Native reserves with the written consent of the Chief and the obligation to pay royalties under Section 28, which shall be used for the benefit of the tribe occupying the Reserve, shall apply to the holders of claims on such grants.

However, there was an entitlement in the Act to allow for public prospecting at that time, but "if the High Commissioner deems it not to be in the public interest to declare any area of land open to public prospecting and mining, he may, subject to any prior rights existing at the date of taking effect of this Proclamation in respect of minerals on such land, confer exclusive rights to prospect and mine thereon by means of a Crown grant".

After obtaining permission to prospect in the late 1930s, it was only in 1954 that DBCM, under the leadership of Dr. Gavin Lamont, eventually commenced preparations to enter the territory to prospect for diamonds. The delay was due to the fact that DBCM had been extremely busy already mining diamonds in South Africa at the De Beers, Kimberley, Dutoitspan, Wesselton, Jagersfontein, Premier and Bultfontein Mines.

However, over the years, perseverance by DBCM and other prospecting companies has, according to the Geological Survey Department of Botswana, now led to the discovery of 287 kimberlites (excluding kimberlites found in active prospecting licences) and two lamproites and by the end of 2013, there shall be no fewer than eight operating diamond mines, four belonging to Debswana, the other to four different companies, namely Boteti Diamonds – Karowe Mine, Firestone Diamonds-B/K11 Mine, Mantle Diamonds (Previously owned by DiamonEx) – Lerala Mine and Gem Diamonds – Ghaghoo Mine. The kimberlites are found in twelve different "Fields" or clusters, the one with most kimberlites being Tshabong Field (85 Kimberlites) followed by the Orapa Field (81 Kimberlites). In November 2011, there were 460 valid diamond prospecting licences in the country, under the management of 52 different companies. These licences cover approximately half of the 581,730 km$^2$ total area of Botswana, the 47th largest country in the world.

Prospecting in Botswana began in earnest in 1955, under the management of Kimberlitic Searches, a subsidiary of De Beers, in the Tuli Block area. Dr. Lamont, recognised universally as the discoverer of the Orapa, Letlhakane, Damtshaa and Jwaneng diamond mines, led the team with visionary leadership skills but was also a very hard task master to the young graduate geologists, Field Officers and other staff under his command. "Doc" as he was so respectfully addressed, sadly passed away four years ago after fighting cancer for some years.

*A/K1 discovery pit - A, left to right, Jim Gibson, Manfred Marx and Gavin Lamont (De Beers, 1967)*

# Prospecting and Evaluation

The following obituary was posted by De Beers, his employer for 33 years, on 21st April, 2008:

## DR GAVIN THOMAS LAMONT

(3rd November 1920 to 19th April 2008)

*"Dr Gavin Lamont, the geologist who led the decade long search for the diamond mines of Botswana, died this weekend in Somerset West. His leadership, tenacity and deep knowledge of geology and of nature contributed to the discovery of the Orapa and Letlhakane diamond mines in 1967 and of the world's richest mine, Jwaneng, in March 1971. These discoveries, together with the prudent management by the government of Botswana of the diamond wealth that flowed from the partnership with De Beers, set the foundation for the economic success of the country and its leading role today in the modern diamond industry.*

*Lamont was born in Kimberley on 3rd November 1920, and was educated at Rondebosch Boys' High School. In 1938 he commenced his studies at the University of Cape Town (UCT). He broke his studies to volunteer for war service in 1939. During the war he distinguished himself in the Special Service Battalion, the long-range Allied desert group operating behind the German lines, until being invalided-out and sent back to UCT to complete his B.Sc. degree, which he did in 1941. From 1942 to 1943 he worked for AECI as a plant superintendent in the cordite plant at Modderfontein Dynamite Factory before furthering his studies at UCT for a M.Sc. degree in geology, received in 1945, and a PhD in 1947. It was at UCT where he met and became friendly with Dr Louis Murray who, on becoming Consulting Geologist to De Beers in the mid-1960's, supported and encouraged Dr Lamont's then ground breaking work in Botswana, leading to the discoveries of the Orapa and Letlhakane diamond mines and, soon afterwards, Jwaneng.*

*These diamond discoveries, which brought about the transformation of Botswana from a mainly cattle-based economy into, today, the biggest and most influential nation in the worldwide diamond industry, did not come about smoothly. They came through the remarkable combination of perseverance, innovative and contrarian thinking and team-building that were the outstanding characteristics of Dr Lamont, or "Doc" as all his staff knew him, and of his all-embracing scientific passion for the geomorphology, botany, zoology, history and archaeology of the areas in which he worked.*

"Doc" had started working in Botswana as a member of the then Bechuanaland Protectorate Department of Geological Survey in 1949. In 1955 he joined Anglo American Corporation, and was seconded to De Beers to conduct diamond prospecting in the Botswana Tuli Block, along the country's eastern border. The river drainage soil sampling results were thin, and with difficulty he persuaded the company to let him look further to the West and North, on the premise that continental crust up-warping could have interrupted the drainage system, and where the weak drainage and extensive thick sand and other Kalahari sediment cover had led others to believe that any economically-worthwhile mineralisation would have been sealed off and made inaccessible. He pioneered and evolved a series of systematic indicator mineral exploration techniques suited to work in the Kalahari, and despite the initial lack of success he persevered with them. He was a firm believer that mines were to be found not by spending time in town or office, but by "boots on the ground", and careful observation. In early 1967 the first kimberlites in the Orapa–Letlhakane area in North-Western Botswana were discovered, initially by means of rapid systematic reconnaissance soil sampling in areas with thinner Kalahari cover, with the Orapa mine, the biggest, being identified in April that year. Orapa is the second-biggest mined diamond pipe in the world.

In the South of the country, on ground previously relinquished under the exacting conditions of tenure of State Grants and later re-awarded to De Beers, Dr Lamont came to postulate that bioturbation, the action by burrowing animals and particularly termites, during past warmer and wetter climatic periods could have brought to the surface kimberlite indicator minerals like ilmenite and garnet from pipes concealed by the sand. Towards the end of 1969, from soil samples taken over 150 square miles, the garnet and ilmenite anomalies associated with the Jwaneng pipe to the west of Lobatse were identified, and the process began of defining its position under some 60 metres of sand cover, and of adapting the relatively primitive drilling equipment of the day to ascertain the presence of diamonds and the possible grade.

Through the innovative and pragmatic use of wider-diameter jumper-drill bits the extraordinary potential of Jwaneng was rapidly established; today it is the richest diamond mine in the world. The modern geophysical and computer technologies of today were not available to "Doc" and his team, but their exploration successes are yet to be surpassed. "Doc" was an outstanding mentor and professional guide to his fellow-workers in the field, and after sometimes initial mixed feelings about his style of management he became a firm friend to many of them. Yet he was at core a private man, self-effacing, of spare frame and almost austere. In writing about the discoveries at Orapa and Jwaneng, he was careful to record the contributions of those who worked with him; Manfred Marx, Jim Gibson, "Brot" Kemang Malema and his brother Eleven and their cousin, Windvoel, Keith Huxham and Guy Lamont,Mike Whately, Bruce Lynn, Peter Bickerstaff, Stuart Vercoe, Norman Lock and Tony Robertson to name only a few. Lamont had a great affinity with the people of Botswana, and in September 1977, in recognition of his contribution to the nation he was awarded the Presidential Order of Honour by President Sir Seretse Khama.

He had accompanied Sir Seretse and Lady Khama, together with Harry Oppenheimer and Barry Hawthorne, on a first aerial overview of the Orapa pipe shortly after its discovery. Orapa was the first diamond mine to be found after eleven long years of perseverance on the ground and in what was then revolutionary thinking. In 1980, after 25 years of service in Botswana, Lamont retired to Somerset West in the Cape, where he settled into retirement with his wife Toni whom he married in 1949. She passed away in 1982. Gavin died peacefully in Somerset West on 19th April 2008, after a long battle with cancer. His son, Guy and stepdaughter Rosemarie both predeceased him. He is survived by Winnie his second wife of 25 years".

# Prospecting and Evaluation

*Geologist Manfred Marx, Orapa (De Beers, 1967)*

Prospecting geologists working for the Consolidated African Selection Trust (CAST) found the first confirmed diamonds in Botswana in 1955 – three small stones along the Motloutse River, near Foley, about 50km south of Francistown. They never found their source and relinquished that licence in favour of De Beers who had been given a Crown Grant to prospect in the Bamangwato tribal territory in 1962. After 1966, the Crown Grant became the State Grant and the Prospecting Licence, and Mining Licence as well, are both now controlled by the Mines and Minerals Act, 1999. According to the Act, the Minister of Minerals, Energy and Water Resources will grant a prospecting licence if -

(a) the applicant has, or has secured access to, adequate financial resources, technical competence and experience to carry on effective prospecting operations;

(b) the proposed programme of prospecting operations is adequate and makes proper provision for environmental protection;

(c) the proposed prospecting area is not the same as, nor does it overlap an existing prospecting area, retention area, mining area or minerals permit area in respect of the same mineral or associated mineral; and

(d) the applicant is not in default.

For the first 25 years of prospecting, geologists relied on tried and tested soil sampling methods to discover kimberlites. The diamond host rock contains a number of trace minerals, such as ruby red garnets, apple green chrome diopsides, jet black ilmenites and pale yellow olivines (collectively named "indicators"), often just a millimetre or two in size, which, when found in the Kalahari sands, brought much excitement back to the old De Beers prospecting bush camps.

*Cr-Pyrope (ruby / purple garnets)*

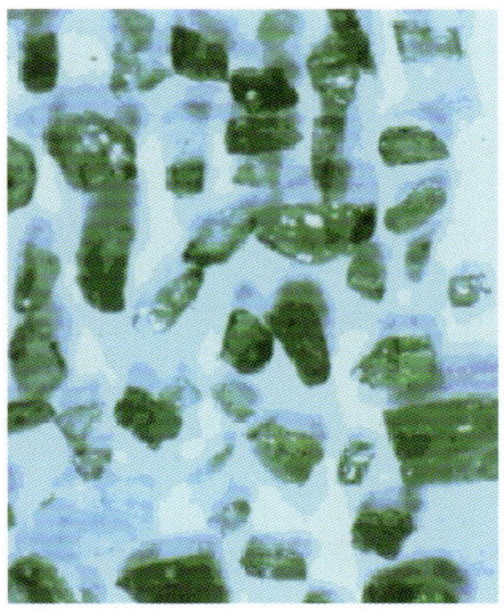

*Cr-diopside (emerald green)*

*Olivine (pale yellow-green)*

*Mg-ilmenite (black)*

*Kimberlite Indicator Minerals*
*Source: Terrain Sciences Division, Geological Survey of Canada*

Dr. Lamont pioneered the soil scooping method of sampling along sampling routes which spanned baselines which were cut through the Kalahari Desert. Some tracks were over 200 km long and have today become well-used "highways" for farmers and campers wanting to explore the more remote parts of the Kalahari. Cutting the tracks through the open savannah grasslands was usually quite easy work with tens of kilometres being completed each day. However, on some occasions, thick Savannah tree vegetation was encountered, and then the teams were lucky to cut one kilometre a day.

*Motswana Team Leader Pono 'Eleven' Malema, Prospecting in the Orapa area (M. Marx, 1967)*

If a prospecting licence measured 50km by 20 km a baseline would be cut through the centre and the desert sands would be sampled on each side. A prospecting team would typically comprise a team of ten people: a team leader or compass man, to keep direction, a bicycle man, to measure distance, three scoopers, to collect the soil samples, four baggers and bag carriers, to sieve, collect, label and carry the soil samples and the geologist, who would follow at the rear and ensure all bags were labelled correctly.

# Prospecting and Evaluation

*De Beers Roads SW Kalahari (R. Fowler, 1983)*

A working day would commence at 5am or before, depending on the time to be taken to reach the pegged baseline. A 20km walk, six days a week, was the order of the day. The team would cut their way through the bush like an aerobatics display team, spearheaded by the Team Leader, flanked by soil scoopers on either side and one immediately behind him. Soil samples were bagged and then sieved, to extract unwanted grasses etc. every half a kilometre until the team returned to the baseline from where they had started, to be met by a driver to return to base camp, which might be up to 50km away.

*Typical Bush Prospecting Camp, Kgalagadi District (M. Brook, 1982)*

*Mike Brook's Debot Prospecting team, Mabuasehube, Kgalagadi District (M. Brook, 1982)*

*Geologist Mike Brook's Mabuasehube Prospecting Camp (N. Muir, 1982)*

*Mosarwa man from Mabuasehube - frequent visitor to Prospecting Camp (M. Brook, 1982)*

After an early lunch and a short siesta, afternoon work would commence with washing and gravitating all the samples collected that day. After gravitation, the heavier, denser grains would be scooped, bagged and labelled to be sent to the mineral laboratory, first in Kimberly, later in Lobatse. Weeks or months later, the "indicators" results per sample were returned to the prospecting geologist to plot on the 0.5 x 0.5 km grid within the licence, and any significant anomalies found after contouring would be followed up with more detailed sampling, basic magnetometer geophysics or drilling. In the areas with thick Kalahari sand cover, there would have been numerous events of soil erosion and deflation which would often disturb the "indicator" distributions, giving false anomalies sometimes, as is the case in the Dutlwe area, northwest of Jwaneng Mine.

# Prospecting and Evaluation

*A typical prospecting camp move in the trustworthy Bedford J5 Truck (De Beers, 1968)*

Whilst it was Dr. Lamont's leadership that directed all the prospecting operations, it was left to hard working and conscientious Batswana, from all over the country, to enact the arduous soil sampling campaigns and ensure samples were properly processed and labelled for dispatch to the laboratory. The "Docs" right hand field officer was Pono "Eleven" Malema, now 78 years old and living in Molalatau in the Central District of Botswana, who controlled all labour in the base camps. He worked for De Beers Prospecting for 35 years between April 1961 and March 1994. Quietly spoken, he commanded so much respect from the De Beers geologists and young Batswana field hands who were fresh and keen to make an impression. His younger brother, "Brot" (means bread in Setswana) Kemang Johannes Malema,

*Gravitating a drill sample from Orapa Kimberlite Field (M. Brook, 2007)*

*Old Prospecting Gravitating Screen with heavy "indicator" minerals at Center (De Beers, 1967)*

mid 1970s, the heyday of kimberlite discovery in Botswana, were Jim Gibson, Manfred Marx, Mike Whateley, Leigh Harrison, Stuart Vercoe, Tony Knowles, John Horner, Jim Platt, Keith Huxham, Jeff Lerwick and Norman Lock.

Geophysical techniques are also often effective in diamond exploration. They can in most cases accurately detect and map kimberlite and lamproite pipes. The most cost-effective and successful geophysical reconnaissance technique used in Botswana has been airborne magnetics, flown by either fixed wing aircraft or helicopter.

*Small diamond at centre of gravitating screen concentrate, Orapa (De Beers, 1967)*

*Diamond prospecting rotating recovery pan (T. Cessford, De Beers, 1968)*

now 77 years old and living in Palapye, equally as competent, managed the gravitating or "jigging crews". "Brot" started his career with De Beers Prospecting in 1955 when he was recruited by "Doc" and retired 40 years later in 1995. He was responsible for managing the washing and gravitating of all the prospecting samples which led to the discovery of the Orapa, Letlhakane and Damtshaa Mines. Other senior Batswana field crew that were also instrumental in helping to find the diamondiferous kimberlites were Rexon Saranyana, Setekia Wasanena and Jacob Ramorwa.

Other prominent geologists that devoted their early twenties and thirties to prospecting life in the Botswana bush between the mid 1950s to the

In 2008, De Beers also experimented with a Zeppelin. With evidence that a major contribution to kimberlite and lamproite magnetic anomalies is often remnant magnetisation, local anomalies may be of normal or reversed polarity compared to a non-magnetic background. It is common that Botswana kimberlites have reversed magnetic polarity.

Airborne electromagnetics (AEM) is also extensively used in diamond exploration. While its cost is a factor of three greater than magnetics, it is particularly effective in detection of weathered or crater facies pipes and is extremely rapid to execute. AEM assists selection of targets for ground follow-up in areas where kimberlite magnetic anomalies are small.

# Prospecting and Evaluation

*De Beers/Debswana Zeppelin used for Geophysics surveys (Debswana, 2005)*

*Orebonye Six Diepo undertaking a Magnetometry traverse in SW Kalahari (R. Fowler, 1983)*

It is also useful to discriminate targets where other sources of localised magnetic anomalies are common. Complementary surface mapping tools (for example airborne radiometrics, aerial photography, digital topography) have all been of assistance in the exploration process.

Gravity, resistivity and electromagnetic surveys, which depend on density differences between the different country rocks of the Botswana craton environments, have also been used to successfully earmark kimberlites for drilling and delineation.

Detailed ground geophysics can usually map out the surface extent of kimberlite or lamproite pipes, and may identify separate intrusions within one kimberlite or lamproite. Since separate intrusions may have very different diamond

grades, this is of significant importance in the location of initial geological sampling. Ground magnetics together with either electromagnetics or resistivity are recommended for detailed mapping and multiple intrusion recognition over a potential kimberlite or lamproite.

Seismic refraction and reflection may provide useful and accurate data but at a significant cost that is not usually justifiable in routine kimberlitic exploration. Determination of geometry, such as subsurface pipe shape or depth to top, may be possible in certain circumstances.

Between 2005 and 2007, De Beers operated a German crewed Zeppelin, one of only three in the world at the time, in the Jwaneng and Orapa Mine lease areas to try to discover new kimberlites using high resolution gravity surveying. The Zeppelin, which weighed 700kg, flew at less than 50m above ground and was successful in finding a small number of pipes. In 2007, however, whilst parked at Jwaneng airstrip, the large Zeppelin broke loose from its mooring and crashed, bringing the project to an end.

In the roughly chronological order in which they have been explored, the twelve kimberlite fields currently existing in Botswana are:-

1) **Kgatleng Field**
2) **Orapa Field**
3) **Jwaneng Field**
4) **Kikao-Khutse Field**
5) **Kokong Field**
6) **Tshabong Field**
7) **Lekgodu Field (Middlepits)**
8) **Gope Field**
9) **Mabuasehube Field**
10) **Martins Drift Field**
11) **Okwa Field**
12) **Nxaunxau Field**

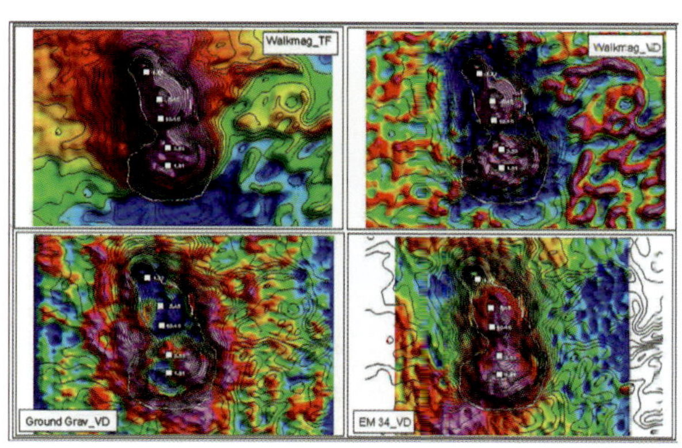


*Ground Magnetic, Gravity and Electro-magnetic survey images over Kimberlite A/K6, Karowe Mine (Lucara, 2010)*

# Prospecting and Evaluation

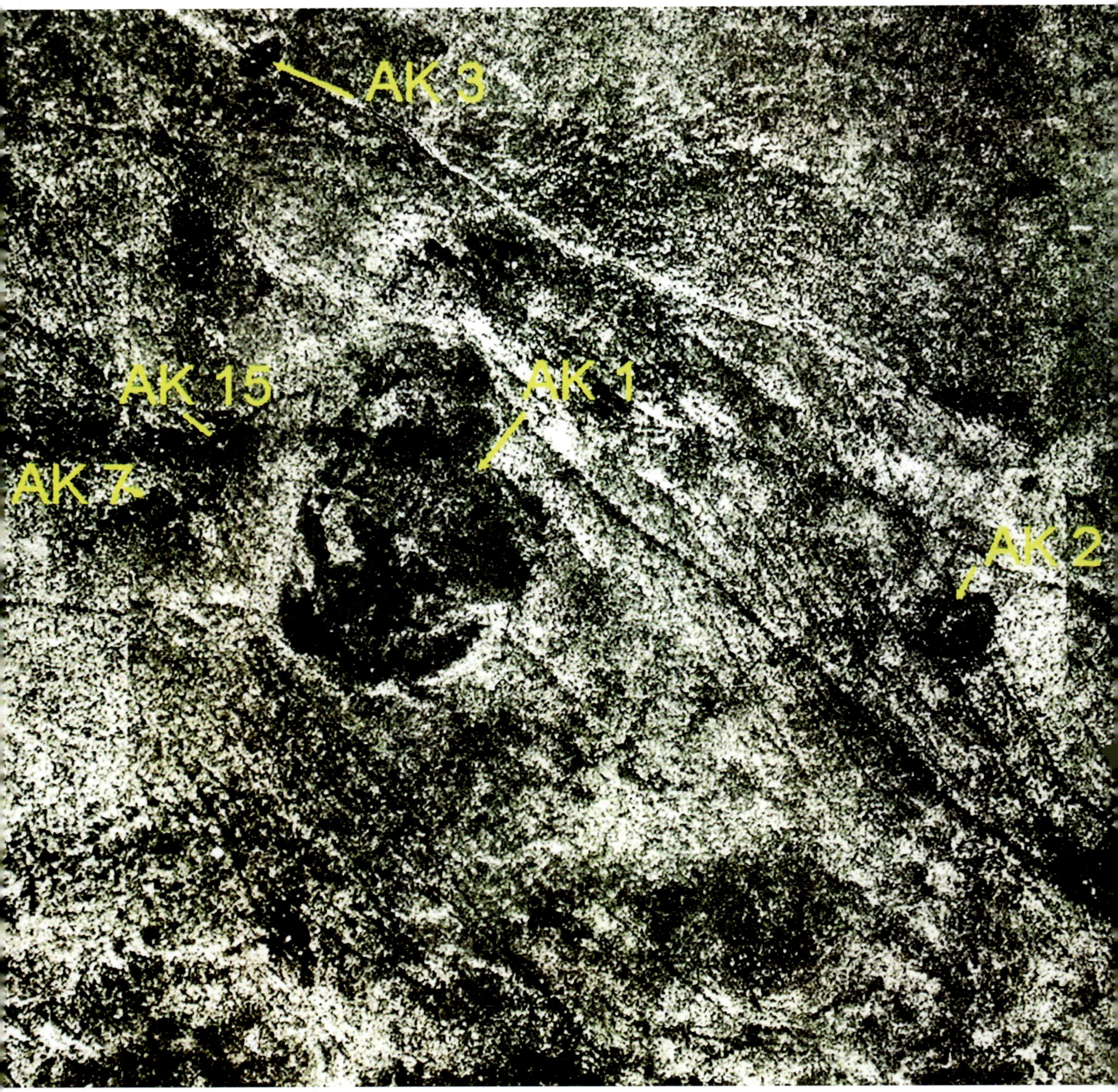


*Photo mosaic of Orapa area, c.1957. The black vegetation anomalies represent more dense bush and enhanced tree growth due to saturated Kalahari sediments overlying less permeable kimberlite (De Beers)*

*Jim Gibson, Orapa Chief Geologist, Debswana 1967 - 2002 (De Beers)*

*Jim Gibson (left) and Manfred Marx at the launch of the Prospecting in Africa Book, 2011 (Courtesy, Jim Gibson)*

*"Although I did not realise it at the time, our discovery of the Orapa kimberlite pipe was to dramatically alter the face of the soon-to-be fledgling democracy of Botswana in so many amazing ways, and also totally determine the course of my life forever".* Jim Gibson, 2012

## KGATLENG FIELD (11 Kimberlites)

After eleven years of prospecting work by De Beers (1955-1966), young De Beers Scottish geologist, Jim Gibson, who was later to become the chief geologist at Orapa Mine for the remainder of his career, found Botswana's first two "kimberlites", Mochudi - 01 and 02 at Morobe and Ramokolonyane near Mochudi in the Kgatleng District. Despite being full of ilmenite indicators, these pipes turned out to be non-diamondiferous lamprophyre kimberlites and no further prospecting in the area took place until the early 1980's, when De Beers revisited the Kgatleng District. Further prospecting by De Beers Prospecting at Sikwane in 1997, discovered a further 9 non-diamondiferous kimberlites. There is no active prospecting currently in this field; Firefox investments were the last to prospect, in 2010.

# Prospecting and Evaluation

## ORAPA FIELD (81 Kimberlites)

In 1966, De Beers, through their lack of success in Botswana, almost pulled out of prospecting in the country. Gavin Lamont pleaded with the then Consulting Diamond Geologist in Johannesburg, Dr. Louis Murray, to allow him and Jim Gibson a few months during the winter of 1966 to undertake reconnaissance sampling in the western Kalahari Desert out towards Orapa.

"Doc" had developed a notion (Wilson, 1971) that the three diamonds, discovered at Foley in 1955, may have a source much further west and that the headwaters of the Motloutse River, which had transported the diamonds, may have been truncated by crustal warping, a feature of Continental Drift, which Dr. Alex Du Toit, the "Doc's" external examiner for his PhD, had previously researched. Dr. Murray approved the trip; thank goodness, because it was this trip that eventually led to the discovery of the Orapa Kimberlite Field in which today the Debswana Orapa, Letlhakane and Damtshaa mines operate, along with Boteti Mining's A/K6 Karowe Mine and Firestone Diamond's B/K11 mine.

In only three months, Lamont and Gibson had covered an area of approximately 4600 square kilometres. Traditional walking methods of sampling along traverse lines could not be followed as it would be too slow. The two geologists set off in two vehicles with a number of Batswana, following tracks made by hunters and cattle farmers throughout the vast area, and collected soil samples at regular intervals. There were no formal roads. At the end of the rapid soil sampling campaign, they established a base camp for washing and screening the samples. To their absolute delight and excitement, they found numerous "indicators" in many samples and knew right away that they were onto the tracks of a kimberlite. During this waiting period their team was expanded by a recent graduate geologist from Cape Town, Manfred Marx. When they resumed exploration in the area, positive results followed rapidly.

*De Beers Geologists Jim Gibson, Dr. Gavin Lamont and Manfred Marx at Orapa A/K1 kimberlite evaluation pit (De Beers, 1967)*

Firstly a five hectare kimberlite, 2125B/K1, was discovered. This became the first diamondiferous kimberlite to be discovered in Botswana, on 1st March 1967, rapidly followed, on 17th March, by 2125B/K2.

Both these kimberlites were characterized by a rise in the kalahari sand surface. Rapid baseline sampling continued in the area and another "indicator" anomaly was explored by Marx when a pit was dug over it. In the afternoon of 19th April, 1967 Manfred Marx climbed out of the pit with the first piece of kimberlite from the world famous 118 ha 2125A/K1 kimberlite - Orapa, soon to become Orapa Mine.

*Dr. Gavin Lamont and geologist Manfred Marx at Orapa A/K1 bulk sampling pit (De Beers, 1968)*

The nomenclature used for naming a kimberlite discovery, e.g. 2125A/K1, is derived from the 1:100,000 Botswana Surveys and Lands topographic map sheet - the first four numbers representing the latitude and longitude of the map sheet, the next letter shows the quadrant location, from top left to bottom right - A,B,C and D and the K1 is the sequential number used to list the kimberlite found within that quadrant.

*De Beers Geologist, Manfred Marx, inspecting A/K1 Orapa kimberlite at an evaluation pit bottom (De Beers, 1967)*

*South African De Beers Geologist, Manfred Marx, holding Orapa A/K1 kimberlite specimen (De Beers, 1967)*

# Prospecting and Evaluation

*Typical working gear above Orapa A/K1 Kimberlite Evaluation pit (De Beers, 1968)*

On November 14th, 1968, the public relations department of De Beers Consolidated Mines, Kimberley, issued for following press release:-

**MAJOR DIAMOND DISCOVERY IN BOTSWANA** - *De Beers has now established that the diamond pipe at Orapa in central Botswana was a major discovery, said Mr. H.F. Oppenheimer, chairman, speaking in Gaborone last night. "This is very important for us, and we also believe that it is important for the new country of Botswana." It seemed rather providential, Mr Oppenheimer continued, that the discoveries which had been made recently – not only by De Beers and Anglo American Corporation, but by other groups too – should correspond with the achievement of independence by Botswana. "I cannot help feeling that somehow providence intends that this should be a highly successful and happy country". Mr Oppenheimer said that he was proud of what the Group's geologists had achieved, adding: "I do hope that our friends here in Botswana will feel, as we do, that the discovery of an important diamond mine is going to make a real contribution to the development of the country."*

***Another published article on the A/K1 Orapa kimberlite discovery from the South African Mining and Engineering Journal (1969), reads…."*** *After discovery, an initial valuation was made by excavating widely-spread pits, the ground being treated in rotary pans. In the present detailed evaluation more closely spaced pits are being sunk to a depth of 120 ft. The material is treated in three heavy media separation plants, with a combined capacity of 300 tons a day. The final diamond-bearing concentrates are passed over vibrating grease belts."*

*Jumper drilling for kimberlite evaluation, 1970s, Jwaneng (De Beers)*

The Orapa kimberlite was named after a nearby "moraka" or cattle post and the mine was officially opened in 1971 at a development cost of 21 million South African Rands. It has an age of 93 Million years and is situated within the Zimbabwe Craton. By the end of 1972 a total of 30 kimberlite discoveries had been made in the Orapa Field, including the Letlhakane 2125D/K1 and K2 pipes. These pipes were brought into production as the Letlhakane Mine in 1977. The D/K1 pipe is characterised by an elevated topography in the form of a calcrete hill adjacent to the Letlhakane River. The mine canteen and quarters were built on the hill. Alex du Toit had already recorded this hill during a mapping exercise in the 1930s. The last kimberlite to be discovered in the Orapa Field during this phase of exploration, 2125B/K16, was also the first kimberlite to be discovered in Botswana with the use of airborne magnetic surveys in 1970. (Daniels, L, 2004).

Between 1966 – 1976, De Beers had discovered a grand total of 79 kimberlites, 75 of which were in the Orapa Kimberlite Field and eight were diamondiferous; this was truly the heyday for them in Botswana. According to Daniels, (Daniels, L 2004) *"the discovery of the Orapa Kimberlite Field changed the De Beers' view of Botswana's exploration potential and their activity increased significantly. Lamont was given a budget to hunt for kimberlites in northern Botswana"*. Just as the first phase of exploration decreased in the Orapa area, activities increased elsewhere leading to the discovery of the Jwaneng Kimberlite Field, approximately 150 km west of the capital city, Gaborone.

Between 1992 and1997, De Beers prospecting returned to the Orapa Field and discovered a further 41 kimberlites.

*Jumper Drilling a kimberlite prospect in the Kgalagadi District (M. Brook, 1982)*

*Bulk diamond sampling at B/K9 using classic jumper rig - to become Damtshaa Mine in 2003 (M. Brook, 1984)*

# Prospecting and Evaluation

*Old diamond core drill rig at Orapa A/K1 (De Beers, 1967)*

Today, Firestone Diamonds also holds prospecting licences in the Orapa Kimberlite Field, in addition to the mining licence for the B/K11 diamond mine. Here they have access to 21 kimberlites, 14 of which are known to be diamondiferous, including B/K 16 which was first sampled by De Beers in the 1990's. The current grade is estimated at 15 cpht (carats per hundred tonnes) and contains high value diamonds, like B/K11. However, it has a larger resources potential than B/K 11 with 17 million tonnes of ore down to a depth of 200m below ground. The detailed evaluation of B/K 16 is expected to commence once B/K 11 mine is in full production by the end of 2012.

*Orapa A/K1 diamond prospecting Plant (De Beers, 1968)*

*Outline Drilling of Orapa A/K1 kimberlite using shallow tractor-mounted air-drill (De Beers, 1967)*

# Prospecting and Evaluation

This field is currently the most active in prospecting, with the following 13 companies holding prospecting licences; Newbright, Seolo Botswana, Kopje, De Beers Botswana Holdings, Firestone Diamonds, Atlas Minerals Botswana, Monak Ventures, Mr/Miss Culture and Heritage World, Longmark Holdings, Lebatla Holdings, Eversharp Investments, Debswana Diamond Company and Competition Area.

## JWANENG FIELD (28 Kimberlites)

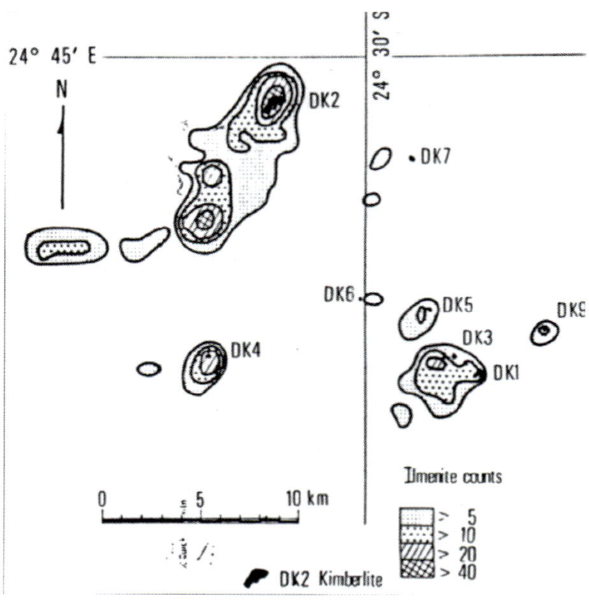


*Contoured Jwaneng Ilmenite Indicator Mineral Map (Lock, 1985)*

*Bulk sampling, Letlhakane Mine tailings by Bauer Technologies, South Africa. 1.5m diameter holes are drilled down to 60m to sample for diamond grade and size (P. Gaegopolwe, 2012)*

According to Lock (1985), by the late 1960's, early 1970's, most of Botswana's Kalahari Desert had been interpreted from using recent regional airborne geophysical surveys and sub-regional ground follow up. These surveys then became the basis for ongoing kimberlite exploration programs.

However, at Jwaneng, the main exploration method was still systematic soil sampling of surface Kalahari sediments, but only a few "indicators", mostly ilmenites, were found. Bioturbation from burrowing termites below termite mounds helped to bring what few "indicators" there were to surface. As seen at the Jwaneng Mine open pit, termites are able to burrow as deep as 70m below ground to search for moisture in the unconsolidated Kalahari sediments.

*Typical Kalahari termite mound - the termites help to bring important "indicator" minerals to surface (M. Brook, 1982)*

De Beers soil sampled over an area of about 750 km$^2$ and found low resolution ilmenite anomalies, which on closer inspection, became more obvious. However, it was subsequently confirmed that the "indicator" anomalies were invariably displaced to the southwest as a result of transport by the prevailing northeast winds. Extensive drilling eventually homed into the pipes. In 1974, after most of the Jwaneng Field kimberlites had already been discovered, 9444 line kilometers of airborne magnetics and gravity geophysics were flown, however, it became evident that few pipes, even D/K2, the current Jwaneng Mine, had a signature. In fact, only 2424 D/K1 and 2424 D/K7 were recognizable from the surveys, the latter being the second ever minable Botswana kimberlite to have been discovered from an airborne geophysics program of exploration.

The five hectare, 250 year old 2424D/K7 pipe was evaluated between 1983 to 1985 by means of sinking a 168m deep shaft. Located 7km south east of Jwaneng Mine, this diamondiferous deposit will be mined by Debswana in the future.

*Headgear over D/K 7 Pipe, 7km south-east of Jwaneng Mine (P. Gomez, 2012)*

The first kimberlite discovered in the Jwaneng area, 2424D/K1, was drilled in 1972. This was followed by the discovery of 2424D/K2 in 1973. The latter kimberlite was approximately 40-45 metres beneath Kalahari formation sediments and consisted of the coalescence of three diatremes in a single crater with a surface area of 54 ha. The kimberlite had a very poor magnetic expression, but was characterised by a distinct gravity 'low'. This 235 million year old kimberlite became known as the Jwaneng Mine which was described by De Beers Chairman, Harry Oppenheimer, at the official opening of the mine in 1982, as *"the most important discovery of the century"*.

*Old Diamond Core Drill, Orapa (De Beers, 1967)*

# Prospecting and Evaluation

*Emptying a bailer sample from large diameter jumper drilling program for kimberlite evaluation at Jwaneng (De Beers, 1970s)*

The Jwaneng Mine comprises three steep-sided kimberlite pipes that connect to each other approximately 100m below present day ground level. The pipes are known as the South, Centre and North Lobes, the centre lobe being the richest in diamonds. These pipes are dominated by two types of kimberlite – a volcaniclastic kimberlite and a dark pyroclastic kimberlite that intrude quartzitic shales of the Transvaal Supergroup. The darker pyroclastic ore is continuous and up to tens of meters in thickness, as opposed to the re-sedimented volcaniclastic ore which is massive and chaotically deposited (Debswana, 2007). The Jwaneng kimberlite is much older than the kimberlites in the other kimberlite fields. A total of 28 kimberlites were discovered in this field.

Currently, this is the second most active diamond prospecting area in Botswana, after Orapa, with the following nine companies still holding prospecting licences; Firestone Diamonds, WCL Investments, Daheng Group Botswana, Silverspring, Waterplus, Geo Perspectives, Debswana Diamond Company, De Beers Botswana Holdings and Amves Enterprises.

## KIKAO-KHUTSE FIELD (23 Kimberlites)

Whilst working at the Botswana Geological Survey Department in Lobatse, Chris Jennings, a South African geologist, developed a model to search for kimberlites using aeromagnetics (magnetic geophysics flown by fixed wing aircraft or helicopter). A review paper published by Pretorius (1978) documented the contribution of the aeromagnetic interpretation to an assessment of the mineral potential of Botswana which attracted a lot of international interest from mining companies.

One of companies, Falconbridge, a Canadian nickel company, decided to employ the aeromagnetics technique and commenced their prospecting campaign in Botswana in 1978.

In 1980, Falconbridge conducted a 78,500 km$^2$ helicopter supported reconnaissance soil sampling program at a sample density of one sample per 165 km$^2$ in the semi-arid Central Kalahari area, part of which is located between Debswana's Jwaneng and Orapa diamond mines. They began investigating areas to the east of Orapa, but were not successful. In 1981, 16 kimberlite pipes were discovered by Falconbridge in this field. They proved to be small and uneconomic and there followed a period of over 15 years of inactivity.

Four kimberlite intrusives, located approximately 15km west-south-west of Kikao, in the southern portion of the Central Kalahari Game Reserve, Botswana, were discovered by MPH Consulting Botswana, operating on behalf of TNK Resources, in early 1997. All four were discovered by a combination of airborne and ground magnetic surveys. The four kimberlites, known as the 173N, 173S, 211 and KO kimberlites are all within 2.5 kilometres of each other. They are overlain by approximately 40m of Kalahari sand cover, and based on core drilling to date, are less than 2 hectares in size.

Kimberlite 173N is circular in outline, and 1.5 ha in size. The 211 kimberlite appears to be elliptical in shape, and between one and half and two hectares in size. All four are diamondiferous. Geologically, the 173N kimberlite is of hypabyssal facies, while 173S, KO and 211 contain both crater and hypabyssal facies kimberlite.

Micro diamonds were found in samples from all four intrusives. Prospecting by the following three companies continues in the area; Zhong Gan Engineering & Construction, Sekaka Diamonds and Eversharp Investments.

# Prospecting and Evaluation

*Modern Diamond Core Drill, evaluating kimberlite deposit, Northern Botswana (M. Brook, 2010)*

## KOKONG FIELD (76 Kimberlites)

In the Kokong Kimberlite Field, some 100km north – east of the Mabuasehube Kimberlite Field, Falconbridge discovered 21 kimberlites between 1978-1986, mostly from airborne magnetic geophysical surveys because the surrounding "country"rocks to the kimberlites were sandstones and shales whose magnetic signature was very quiet. These kimberlite magnetic anomalies were therefore very easily recognisable, even though the Kalahari sediment cover sometimes exceeds 150m in thickness. Discoveries by other prospecting companies increased the number of Kokong kimberlites to 34. The most interesting Falconbridge find was the KN 70 kimberlite with a possible grade of 10 cpht. Although most of the kimberlites had some crater facies associated with them, only hypabyssal facies kimberlite was intersected at the KE1 kimberlite. This kimberlite was also the only one associated with a gravity 'high'. New discoveries by Firestone Diamonds in the last five years has now increased the number of kimberlites in this field to 76.

Later, a joint venture between Trivalence Mining Corporation and Rio Tinto Mining and Exploration Ltd. recovered a number of small diamonds from a kimberlite pipe in the Kokong Joint Venture Project. The joint venture agreement allowed Rio Tinto to fund and manage the diamond exploration program. The majority of these pipes were tested with core drilling. Material from these holes was submitted to Rio Tinto's Polokwane diamond laboratory for recovery of indicator minerals and diamonds.

Material from kimberlite K295 returned 19 small diamonds. The K295 is a large conductive anomaly defined through an airborne Time Domain Electromagnetic Survey flown in late 2004. A total of 18 diamondiferous kimberlties occur in the Kokong Field.

The following seven companies continue to prospect in the area: Daheng Group Botswana, Architect Enterprises, Meifo, Waterplus, Homeland Properties, Firestone Diamonds and Preston Investments.

## TSHABONG FIELD (86 Kimberlites)

In the Tshabong area, prospected by the same joint venture of Falconbridge and Superior Oil that found Ghaghoo Mine, 86 kimberlites, of which 20 are diamondiferous, have been discovered. Most discoveries were made in 1978, when the main target was identified as "M1", being the first anomaly identified in the Molopo project area. Ground crews were sent to locate the target on the ground, which was even larger than the Orapa kimberlite pipe, in excess of 1.4 square kilometres, equivalent to the size of about fourteen English football pitches. An old jumper drill, or "nodding donkey", of the "Prinsloo Harvester" type, was sent out from Gaborone and kimberlitic sediments were discovered at a depth of 80 metres below ground level. After two years of delineation drilling the joint largest recognised kimberlite in the world (the Williamson pipe in Tanzania is also 146 ha), the 77 Million year old, 146 hectare M1 kimberlite was earmarked for bulk sampling. In addition to the M1 pipe, Falconbridge under the guidance of Dr. Chris Jennings, former Deputy Director of the Botswana Geological Survey Department, discovered another 39 kimberlites in the area including the three part T6 kimberlite of 54 ha, which is similar in size to the Jwaneng kimberlite.

The first Falconbridge Tshabong discoveries of the 1970's have subsequently been followed up by Firestone Diamonds from 2004 when they conducted 50 km$^2$ of ground geophysics and 22 km of various types of exploration drilling, with two tonnes of micro diamond samples being taken and a total of six bulk samples.

Macro diamonds (>0.5mm in size) have also been recovered from five kimberlites and the future prospects for economic discoveries are good. Firestone Diamonds, along with two other prospecting companies, Pangolin Gold and Eversharp, continue to hold prospecting licences in this field.

## LEKGODU FIELD (MIDDLEPITS – 8 Kimberlites)

De Beers Prospecting were the first to be active in the Middlepits area back in the early 1970's and 1980's with typical soil sampling and ground geophysics exploration. De Beers conducted preliminary tests on 22 sites in a 250 square kilometer area and reported the presence of a fairly homogenous distribution of diamonds in the gravels present with an average thickness of 3.5 meters. Also, 4.43 carats from 62 diamonds (all diamonds greater than 1 mm in size) were recovered from 68.2 metric tonnes of material. At least one of the diamonds was greater than 0.55 carats in weight. Prospecting in the area then ceased for many years.

The Middlepits Prospect straddles a major north-east trending structure which hosts the Jwaneng and Venetia (South Africa) diamond mines and is localized between two diamond-bearing kimberlite swarms.

In the 1990's, TNK Resources, from Canada, prospected the Middlepits area next to the De Beers prospect and conducted a large-bore drill bulk sampling program on two pipes, adding a large capacity processing plant. The Middlepits Prospect was found to have "diamondiferous gravels reclassified as a "diamond resource" as a result of TNKR's efforts.

In 1995, Layfield Resources undertook a low-level, high-resolution airborne magnetic survey on the company's 860,000 acre Molopo Project.

# Prospecting and Evaluation

*Bulk sampling from large diameter kimberlite reverse circulation drilling by De Wet Drilling - Northern Botswana (M. Brook, 2007)*

The survey was flown by World Geoscience Inc. and entailed some 5000 line kilometers of coverage. The survey was concentrated in the southwest portion of the property in an area partially covering and adjacent to the Middlepits diamondiferous gravels, but a kimberlite bedrock source has never been found.

There has never been any drilling in this critical area of the Molopo property and the discovery of kimberlites here would indicate the presence of a new kimberlite field, a pipe or pipes which could be the source of the diamonds in the gravels.

Layfield then went on to drill 2250 meters of reverse circulation drilling to test priority targets located by the geophysical surveying.

In 1995, Southern Africa Minerals Corporation drilled eight kimberlites in the field. Diamond prospecting continues in the area through Thermophylae Holdings.

## GOPE FIELD (9 Kimberlites)

Between 1978 and 1986, a Canadian company, Falconbridge, had discovered no fewer than 79 kimberlites in Botswana, 19 of which were diamondiferous, the most successful discovery being the 10.3 hectare, economical, Gope 25 kimberlite, which is currently being developed as Ghaghoo Diamond Mine by Gem Diamonds. The pipe was discovered under a Kalahari sand cover of 80m, from a small number of "Indicators" and aeromagnetics. This was an incredible achievement in the space of just eight years.

The Gope kimberlite field is situated in the eastern part of the Central Kalahari Game Reserve and Falconbridge found the first discoveries in the early 1980's when a helicopter based aeromagntic survey was undertaken and many prospective kimberlite targets were identified, including the Gope 25 pipe which was discovered in 1981. Eight heavy mineral "indicator" anomalies were detected in soil samples. Anomaly GO-1 was assigned a high priority, despite being represented by only four relatively small ilmenite grains. The relatively small number of ilmenites recovered at Anomaly GO-1, and their generally small grain-size, are partly attributable to the fine-grained nature of the soil in the Gope area relative to most of the area sampled.

In 1981, Falconbridge's drilling within Anomaly GO-1intersected the 10.3 hectare diamondiferous Gope 25 kimberlite pipe beneath 80m of sand. Eight additional kimberlites were discovered, but only Gope 25 was economic. In 1983, Falconbridge formed a joint venture with De Beers and in 1985, the Gope Exploration Company was formed and in 1988 an exploration shaft was sunk to a depth of 150m, with 200m of tunnels. There followed seven years of inactivity until 1995 when additional holes were drilled and tunnels extended 1.4km, recovering 18,000 tonnes of kimberlite.

In 2007, Gem Diamonds acquired 100% of Gope kimberlite 25 and began preparations for developing a mine under the new name Ghaghoo diamond mine.

Prospecting in this field continues through the Daheng Group Botswana and Sekaka Diamonds.

# Prospecting and Evaluation

*24-inch Reverse Flood Air Assist Evaluation Drilling by De Wet Drilling for the Kimberlite Kx36 Mini Bulk Sampling Program (Petra Diamonds, Central Kalahari, 2012)*

## MABUASEHUBE FIELD (9 Kimberlites)

Between 1980 and 1984, De Beers Prospecting investigated the area between Tshabong and Kokong and they discovered a cluster of six non-diamondiferous kimberlites or lamproites at Mabuasehube. I was one of the De Beers prospecting bush geologists in the area and recall the shear disappointment after the drilling results came through.

In fact, this event contributed to my decision to resign my position to study hydrogeology – I always knew finding water, although not that straight forward in some parts of Botswana, was going to be significantly easier than finding a diamondiferous kimberlite!

Chablis Enterprises (Botswana) Pty Ltd ended prospecting in this field in 2010 and there is currently no active prospecting.

*Diamond Prospecting sampling baseline across Letsibogo Pan, Mabuasehube, Kgalagadi District (M. Brook, 1982)*

## MARTINS DRIFT FIELD (8 Kimberlites)

The Martins Drift or Lerala kimberlites were first discovered by De Beers in 1991. Five pipes, K002 to K006, were discovered with a combined size of 6.33 hectares. However, it was only in 1997 that trial mining commenced. This historical diamond production was known as the Lerala Diamond Project when De Beers Prospecting Botswana operated the "Tswapong" trial mine between April 1997 and December 2000. 381,000 tonnes of kimberlite ore was mined from the five pipes but the grades obtained were uneconomic according to De Beers, and the project was abandoned and the prospecting licence relinquished.

In 2002, DiamonEx Botswana took out their first prospecting licence on the Lerala kimberlites covering an area of 71 km$^2$ and began to reassess the work of De Beers and also drilled additional small core and large diameter evaluation boreholes into four of the five pipes.

In 2005, pitting into the kimberlite followed to allow for a bulk sampling program which recovered 851 carats from 5,476 tonnes of ore treated. The grade was found to be economic and preparations for mining the Lerala Mine by DiamonEx began in 2008.

Prospecting continues in this field through De Beers Botswana Holdings, Daberg Group Botswana and Lebatla Holdings.

## OKWA FIELD (8 Kimberlites)

A detailed airborne magnetic survey was flown over the entire Okwa area by De Beers in the late 1980's in the search for kimberlites. Nothing of significance was discovered and a long period of inactivity followed. In 1990, De Beers returned and discovered seven kimberlites, OKWA -01 to OKWA -07, following up earlier geophysics targets. One additional kimberlite, OKWA-08, was discovered in 1995. None were found to be economic and De Beers relinquished the prospecting licence.

In 1995, Diamondfields Botswana, conducted extensive filtering on the De Beers airborne survey data which then focused attention on a number of additional interesting targets. An important aspect to the Okwa Field Project is that most targets are at very shallow depths, in the order of tens of metres, indeed, Okwa rocks are exposed in the Okwa River Valley and can be prospected from surface.

Botswana Diamondfields conducted a 4 700 line-km helicopter aeromagnetic survey of the southern isolated licence block in November 1995. Detailed airborne follow-up surveys were conducted over five targets identified in the field and another six targets located in the main northern block. An additional three targets were surveyed on the ground. None of the targets were kimberlites.

There is currently no active prospecting in this field. Chablis Enterprises were the last prospector in 2010.

# Prospecting and Evaluation

*The first double cab 4x4 in Botswana - owner, Geology Department, Orapa (M. Brook, 1984)*

## NXAUNXAU FIELD (27 Kimberlites)

To date, Tsodilo Resources Limited has identified 27 kimberlites in the Nxaunxau Field in the Ngamiland District which stretches over an area of some 40km by 25km in a NW-SE orientation. An analysis of the technical characteristics of the known kimberlites in the Nxaunxau field was initiated in 2010. Additional mineral chemistry and petrography work was necessary for some of the kimberlites and this was completed in late 2010. A review of this data resulted in the decision to submit samples from three kimberlites (K4, K10 and K20) for micro-diamond analysis. Approximately 200kg of drill core from each of K4, K10 and K20 were submitted for analyses in Canada in 2010 and two of the three kimberlites submitted for analysis had positive micro-diamond recoveries as follows: the most interesting result was obtained from kimberlite K10 which returned 14 diamonds from 221.2 kg or 63.3 stones per ton (SPT). This is the first time that such a large number of micro-diamonds had been returned from any of the kimberlites in Ngamiland. The recovered diamonds appear to be colourless and fragmented; seven kimberlites out of a cluster of 27 intrusions have been tested for micro-diamonds. Of these, four tested positive for micro-diamonds with one, K10, significantly so. These recent results in combination with previous micro-diamond analyses, indicate that this kimberlite field is clearly diamondiferous.

*Small air rig exploring Orapa pan gravels (M. Brook, 1984)*

The micro-diamond results and the tectonic reconstruction studies continue to enhance the potential of the Ngamiland diamond project and the diamond exploration program in 2011 focussed on kimberlites K17, K18, and K19 which are the most northerly bodies so far discovered in the Nxaunxau field.

Prospecting continues through two companies; Newdico and Eversharp Investments.

The heyday of kimberlite prospecting in Botswana took place between 1966 and 1984, during which period the kimberlite pipes belonging to all of the current eight diamond mines of Botswana were first discovered.

Finding more kimberlites today is becoming increasingly more difficult, but with the current life of mines expected to end within the next 20 years or so, it is essential that all future possibilities of promising finds are exhausted by continuing with the current prospecting campaigns across the whole of Botswana.

Nevertheless, according to John Teeling in March 2012, executive chairperson of Botswana Diamonds, *"The future outlook for the diamond industry remains robust. We do not expect any major new supply to hit the market in the next 10 years"*.

Having understood the various prospecting and ore evaluation methods, lets now proceed to the mines where the diamonds are ultimately recovered.

BUCYRUS
930E
KOMATSU
614

# 5 Mining

Exploration and Mining in Botswana is governed by the Mines and Minerals Act, Chapter 66:01 of 1999 and all diamond mines and places where diamonds are sorted, valued, cut and polished fall under the Precious and Semi-Precious Stones (Protection) Act (Cap 66:03).

The main legislation governs the ownership of minerals and mineral rights (rights are vested in the Republic of Botswana) applications for mining licences and environmental obligations. The Act also provides for financial aspects such as royalties to be paid, licence fees and penalties. Once a mining licence has been issued, the Government has the option of acquiring up to 15% working interest participation in the proposed mine. This provision, however, does not apply to the mining of diamonds where the ownership level is a matter for negotiation. The holder of a mineral concession for diamonds and other precious stones pays a tax of 10% of the gross market value of what has been mined.

# Mining

*1st July 1971, Orapa Mine Officially opened by Sir Seretse Khama, President of Botswana and Harry Oppenheimer, Chairman of De Beers. (De Beers, 1971)*

Diamond mining in Botswana commenced on 1st July, 1971, with the official opening of Orapa Mine, 240km west of Francistown, by President Sir Sereste Khama, in the presence of Harry Oppenheimer from De Beers and John Richards, General Manager of Orapa Mine from 1970 to 1973. He succeeded C.B (Gus) Edwards who had managed the bulk sampling plant from 1967 to 1970. The latest diamond mine to be commissioned is Boteti Diamond's A/K6 Karowe Mine, some 10km west of the Debswana Letlhakane Mine which was put into operation in 2012.

There are currently seven operating diamond mines in Botswana, in order of opening; Orapa, Letlhakane, Jwaneng, Damtshaa, Lerala (formerly Tswapong Mine), B/K11 and A/K6. An eighth diamond mine, Ghaghoo Mine (previously called Gope Mine) is scheduled to commence production in late 2013. The operating mines, mining about 241 hectares of kimberlite, currently typically produce about 34 million carats per year with an additional 600,000 carats being produced once Ghaghoo Mine is commissioned. This combined Botswana production represents about 37% of the global diamond production.

## Botswana's Diamond Mines

| Mine (oldest first) | Latitude | Longitude | Size (Hectares) | Typical annual production (carats) | Life of Open pit Mine (year) |
|---|---|---|---|---|---|
| Orapa (1971) A/K1 | 25 22 08.73 | 21 18 30.63 | 118 | 14,000,000 | Cut 4 to 2030, thereafter underground? |
| Letlhakane (1977) D/K1 & D/K2 | 25 41 22.01 | 21 31 12.24 | 15.2 | 1,000,000 | To 2014. thereafter 20 years tailings production |
| Jwaneng (1982) D/K2 | 24 42 07.12 | 24 31 25.77 | 54 | 12,000,000 | Cut 9 to 2029, thereafter underground? |
| Damtshaa (2003) B/K9, B/K12, B/K1, B/K15 | 25 31 57.67 | 21 18 29.04 | 24.5 | 230,000 | B/K12 to 2017 B/K9 to 2026 |
| Lerala (2008) K002,003,004, 005,K006 | 27 45 21.78 | 22 46 06.65 | 6.33 | 330,000 | To 2021 |
| B/K 11 (2010) | 25 30 28.25 | 21 28 13.77 | 8 | 225,000 | To 2020 |
| A/K 6 Karowe Mine (2012) | 25 28 14.00 | 21 29 59.00 | 4,2 to 7 at depth | 400,000+ | To 2022 |
| Ghaghoo (formerly Gope - GO-25) 2013 - underground mine | 24 46 40.64 | 22 37 14.12 | 10.3 | 600,000 | To 2038 |

# Mining

*Orapa Mine with Orapa Town in background (Debswana, 2011)*

Botswana is now the world's largest diamond producer by both volume and value of output and also hosts the world's single richest diamond mine, Jwaneng, which opened in 1982. Revenue from Botswana's diamond mines currently contributes to about 35% of Gross Domestic Product and over 60% of revenue to Government.

Before describing the individual mines in detail, we provide an introductory view of how diamond mining is actually performed. Mining commences after completion of all bulk sampling activities and an economic diamond grade for the kimberlite has been proven. In all cases, Botswana's diamonds are mined from a primary, in – situ kimberlite and all current operations are open pit or open cast. No alluvial diamonds are mined in Botswana, however, for the first years of mining at Orapa, the diamonds were recovered from a sedimentary, layered sequence, deposited within a lake, which has been eroded and re-worked from within a volcanic crater. In 2013, Ghaghoo Mine will become Botswana's first underground mine, extracting kimbertite ore from a depth of about 150m below ground level.

Once the current open pits reach a certain depth, there is also potential for them to become underground mines. However, mining underground is a lot more expensive and generally recovers a lot less kimberlite ore than open pit mines.

Diamond production in Botswana has increased year on year, with the exception of two periods, 1990 to 1993 and 2008 - 2009, because of a planned reduction in output to match lower sales caused by global financial crises. In 1967, production amounted to 1000 carats only (from the Orapa Mine bulk sampling plant), but rose to a maximum of 34.4 million carats in 2006.

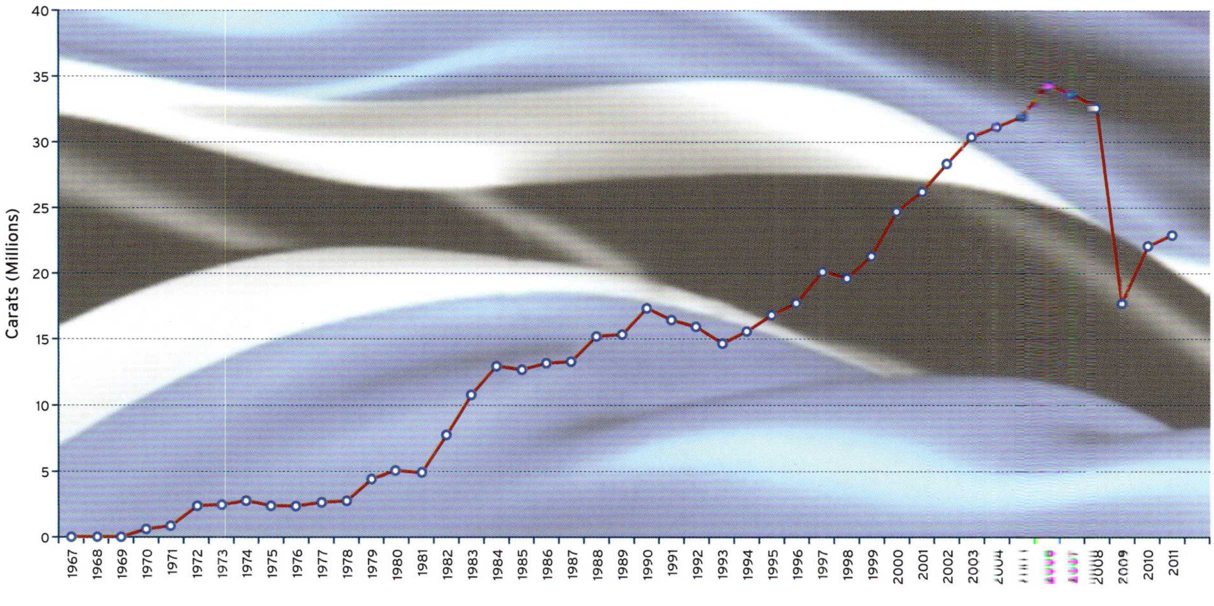


*Botswana Diamond Production 1967-2011 (Data Source: Central Statistics Office)*

# Mining

*Debswana Head Office, The Mall, Gaborone (Debswana, 2002)*

In 2010, Debswana production increased to 22.2 million carats from 17.7 million in 2009. Carats sold were 23.9 million, with total sales of P18.2bn ($2.68bn). The revenue improvement was a result of the increase in prices after the 2009 downturn. In 2011, Debswana produced 22.9 million carats from 86.8 million tonnes of ore and waste mined and received a record $US 3.43 billion in revenue, with an average price per carat of US$ 177.

Louis Goodwill Nchindo was the first and longest serving Motswana Managing Director at Debswana, from 1998 to 2004. During his tenure, great emphasis was placed on local empowerment and citizen training and education and in November, 2001, Leonard Makwinja was appointed General Manager at Orapa and Letlhakane Mines, becoming the first Motswana General Manager at any Debswana operation. In 2003, the current Botswana based Debswana Technical Services operation was established.

In diamond mining, the mine planning process involves the optimal exploitation of the kimberlite resource which is critical and includes:

- Open Pit Optimisation modeling
- The Pit / Waste dump designs
- Production scheduling/sequencing
- Trade-off studies
- Earthmoving equipment forecasts
- Monitoring and Reconciliation - Mine to Plan Adherence
- Mine Strategic Business Plan

*Debswana Technical Support Center - Atrium (M. Brook, 2010)*

*Debswana Technical Support Center, Gaborone, 2010, renamed Debswana Corporate Centre in 2012. (M. Brook, 2010)*

# Mining

*Waste stripping at Jwaneng Cut 8, conducted by Majwe Mining, a JV between Basil Read Mining, Bothakga Burrow Botswana and Leighton International Ltd. (Debswana, 2011)*

*Debswana Jwaneng Mine Main plant, seen from first benches of Cut 8 (Debswana, 2011)*

The mine plan involves designing "cuts" into the ground, which are developed in "benches", usually 12-16m in height. The slope angle of the cut depends on the strength of the country rock mass – the stronger country rocks, e.g. Basalt, at the Debswana northern mines, may have slopes of up to 60 degrees, whereas the weaker rocks, such as mudstones, have much gentler slopes, of the order of 30 - 45 degrees. Since kimberlite diatremes are carrot in shape with the resource tapering at depth, there is usually a requirement to mine several successive cuts before reaching the end of open pit mine life, before considering underground mining, if further cuts are uneconomical. A new mine is accurately surveyed and a production plan put in place which commences with production blast hole drilling at between 4 - 12 inches diameter. Benches are blasted usually one at a time by injecting liquid explosives into the bottom of the blast holes and dynamiting a sequence of holes in a single blast. Explosives are injected at predetermined quantities depending on the fragmentation required.

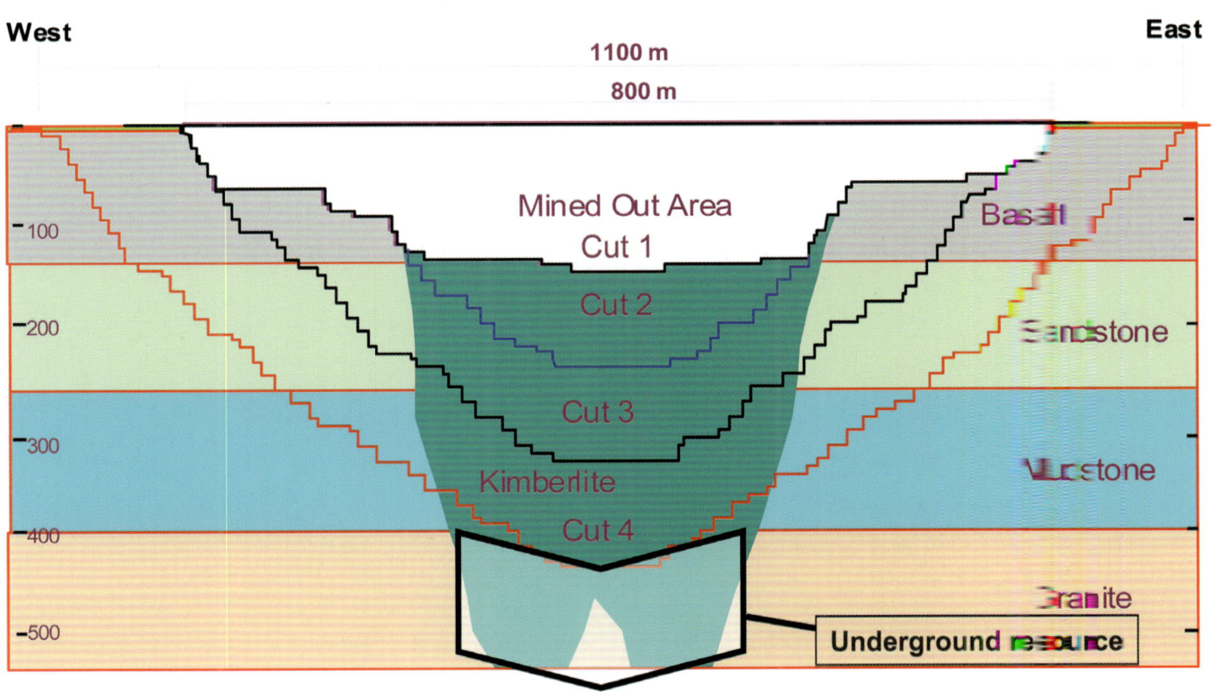


*Cross section of D/K1 Letlhakane Mine showing the four mined "Cuts", Letlhakane Mine designs (Debswana)*

# Mining

*Afican Explosives Limited (AEL) liquid explosive being delivered to blast holes at Jwaneng Mine (Debswana)*

*AEL Liquid explosive being pumped into blast holes at Jwaneng Mine (Debswana)*

The next stage of mining is loading and hauling the kimberlite ore and waste rock to surface. The kimberlite ore is sent to the diamond processing plant and the country rock waste is dumped on waste dumps. If there is excess ore at the time, that ore can be dumped onto low - grade and high grade "stock or ore - piles".

Shovels as well as Front End Loaders are used to load trucks, with bucket capacities ranging from 10 - 38m$^3$. However, for the Cut 8 project in Jwaneng, a new Bucyrus shovel was procured with a bucket capacity of 46m$^3$. Both mechanical and electric shovels are used and on-board electronic systems assist the operator in determining the load before deposition on to the trucks. Loading then takes place into large trucks, which at Debswana, have capacities of between 35 - 300 tonnes.

*A blast at Orapa Mine (De Beers, 1975)*

*Blast Hole Drilling at Debswana Operations (Debswana)*

# Mining

*Truck loaded with 200 tonnes of kimberlite ore at Jwaneng Mine (Debswana)*

*New Jwaneng Cut 8 Super Komatsu 300t truck bodies patiently awaiting their trays to be fitted. A total of 41 trucks are to be procured to load 700Mt of waste (Debswana, 2011)*

*New Jwaneng Cut 8 Super Komatsu 300t truck with tray fitted. (Debswana, 2011)*

GPS controlled trucks are used to haul material out of the mine to various destinations (waste dumps, plants, stockpiles). The maximum truck speed is 15 km/hr. up an 8% gradient ramp when loaded. These trucks are massive and dwarf motor vehicles and people. They make their way up the mine slopes on haul roads before emptying their loads into the primary crusher, which can be located within the pit or at surface. Loading and hauling is a 24-hour business, conducted usually in three shifts.

# Mining

*In – Pit Vehicle Monitoring Control Room, Letlhakane Mine (Debswana)*

All in-pit activities are monitored and recorded by an in-pit control room to ensure that safety remains the number one priority by all for all. Let's take a closer look at Botswana's eight diamond mines, commencing with Debswana's four mines, which currently produce 96% of total diamond output in Botswana. Debswana is the world's leading producer of gem diamonds.

The first years production for Debswana's individual diamond mines, were as follows:

| | | | | |
|---|---|---|---|---|
| 1) | Orapa | (1971) | 1,438,168 | carats |
| 2) | Letlhakane | (1977) | 346,332 | carats |
| 3) | Jwaneng | (1982) | 2,614,442 | carats |
| 4) | Damtshaa | (2004) | 336,299 | carats |

*Groundprobe slope stability radar being controlled by Felix Ramsden at Orapa Mine (Debswana, 2006)*

## ORAPA MINE ("RESTING PLACE FOR LIONS")

Orapa is the largest conventional open pit diamond mine in the world and the world's largest diamond producer by volume and is situated 240 km west of Francistown. The mine began production in July 1971 and is the oldest of Debswana's operations. The mine had to be linked to Francistown by a new road and power supply. Fortunately, ample water supply was available locally from the regional groundwater system comprising the Ntane sandstone aquifer.

A new town also had to be built from scratch; the town originally had 162 houses for married employees and single quarters for 210 men, a hospital, clinic, two schools, an adult education and training center and full recreational facilities. Today, the town's population is about 9,500 and both Orapa and Letlhakane have expanded significantly.

*Dr. Adrian Gale, General Manager, Orapa, Letlhakane and Damtshaa Mines (Debswana 2012)*

*Orapa Mine Open Pit (Debswana, 2006)*

# Mining

*Orapa Mine Main Treatment Plant (Botswana Press Agency, 1980)*

*Orapa Town (Botswana Press Agency, 1980)*

*Orapa Mine aerial view (Botswana Press Agency, 1980)*

# Mining

*Orapa Technical Support Centre (M. Brook, 2008)*

Orapa's production continues to contribute significantly to Debswana's total carat output, in combination with Letlhakane and Damtshaa Mines, 10,7 million carats were produced during 2010 and 11.16 million carats in 2011.

When Edward Epstein (1982) visited Orapa with a De Beers delegation on 5th December 1978, he later wrote, *"I had imagined a mine deep underground honeycombed with labyrinthine tunnels. Instead I saw an open pit that looked like an excavation site for a skyscraper. A number of dirt roads wound their way down to the bottom of the pit, which was no more than 690 feet below the surface of the earth at its deepest point.*

*The Orapa Diamond Mine is featured on the reverse of the twenty Pula Botswana Bank Note. The "Pula" replaced the South African Rand as Botswana's currency in 1976. (M. Brook, 2012)*

*Sebeltela Sebeltela General Manager, Jwaneng Mine 2003 - 2005 and Orapa and Letlhakane Mine 2005 - 2008 and Len Makwinja, appointed first citizen General Manager Orapa and Letlhakane Mine in 1999 and Debswana Deputy Managing Director 2005 - 2007 (Debswana)*

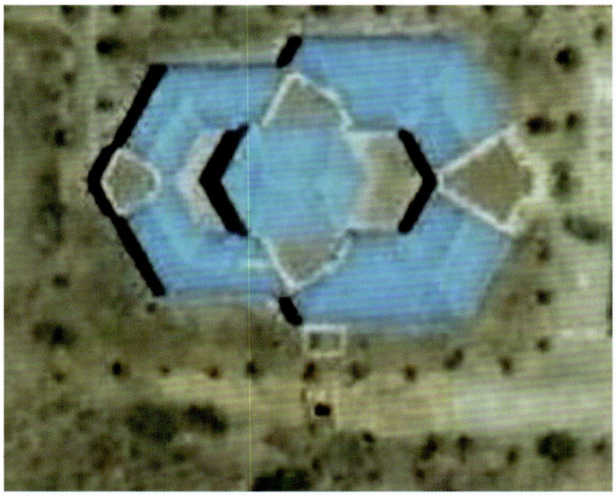

*Google Earth view of the Orapa Mine Technical Office Block in the shape of the Debswana Logo (Google Earth, 2012)*

*Orapa Mine Recovery Plant (De Beers, 1971)*

# Mining

*On the floor of the mine I could see about fifty Botswana workers. They were dressed in khaki jumpsuits and yellow helmets, and most of them were operating steam shovels and other mechanised equipment".*

Justification for commencing mining at Orapa was arrived at from bulk sampling shallow pits across the whole kimberlite pipe, A/K1. A bulk sampling plant was constructed on site in 1968 and the kimberlite ore grade was evaluated to be about 50 cpht but now sits at 80 cpht. Today, the area of excavation at surface measures about 120 hectares and the 205m deep pit measures 1.2 by 1.8km. The planned life of the current open pit is to the year 2030. But this does not include processing of the tailings dumps which have significant potential still for further diamond recovery.

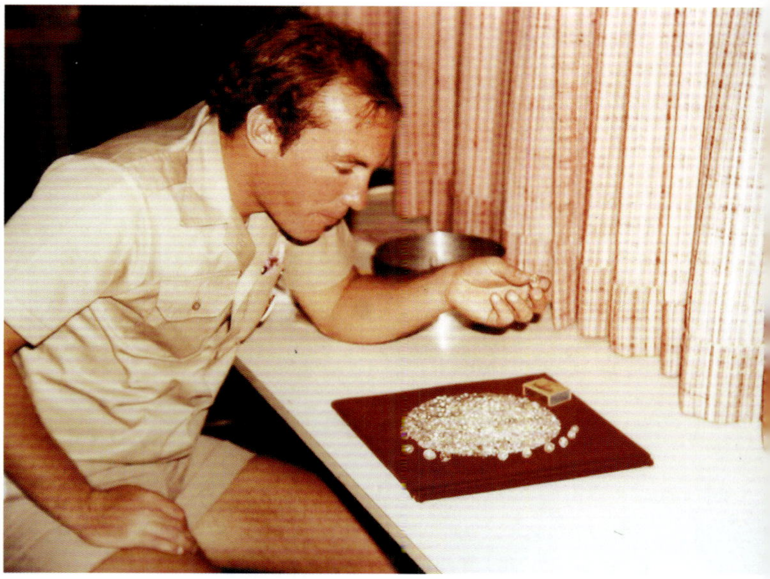

*Mike Brook, geologist Orapa Mine (Debswana 1984)*

*Orapa Mine A/K1 Pilot Plant and Camp (De Beers, 1969)*

# Mining

*Sir Lynden Oscar Pindling, President of the Bahamas for 25 years, visiting Orapa Mine (Botswana Press Agency, 1980)*

*Orapa A/K1 pit trenching and Main Plant (De Beers, 1970)*

# Mining

*Orapa Mine (G. Beale, 2012)*

In August 1996, representatives of Botswana Government, De Beers Centenary AG and Debswana Diamond Company signed an agreement to double production at the Orapa Mine. The expansion increased Orapa Mine's annual production of 6 million carats to 12 million carats from the No. 1 and No. 2 plants from the year 2000 and raised Debswana's total production to about 26 million carats per year.

In years 2005 and 2006, Orapa produced its maximum carats at 15.6 million per year. Besides the Orapa 2000 project, Orapa is committed to its environmental responsibilities and focuses on water conservation, waste management as well as maintenance of the Orapa Game Park.

*Orapa 2000 Expansion - Official Opening of Number 2 Plant - with the latest technology (Debswana, 2000)*

*Orapa Mine (Debswana, 1982)*

*Orapa A/K1 pilot plant (De Beers, 1968)*

*Zebra at Orapa Mine Game Park (Debswana, 2010)*

## Mining

Letlhakane Mine (Debswana, 2006)

## LETLHAKANE MINE ("LITTLE REED")

Letlhakane Mine is situated 40km from Debswana's Orapa operation and 190 km West of Francistown, in central Botswana. The mine which was first discovered during the sampling and evaluation process at Orapa, became Debswana's second mine when it opened in 1977. The mine is managed from the larger Orapa operation.

Two kimberlite pipes are mined, D/K1 (13 hectares) and D/K2 (2.2 hectares). The D/K1 pipe has a "trouser leg" morphology and the area of excavation on surface measures 88 hectares. The current pit depth below ground level is about 350m and the pit dimension is 1.1 x 1.1km. Ore grade is about 30 cpht and the life of the current open pit is to 2014, after which there is potential for additional mining underground or by a vertical mini-pit. In 2011, Letlhakane Mine produced 1 million carats.

In 2007, a mining contract was awarded to a Joint Venture comprising Basil Read Mining and Bothakga Burrow Botswana to mine 13 million tones of ore and waste from D/K2.

*Letlhakane Mine (Debswana, 1982)*

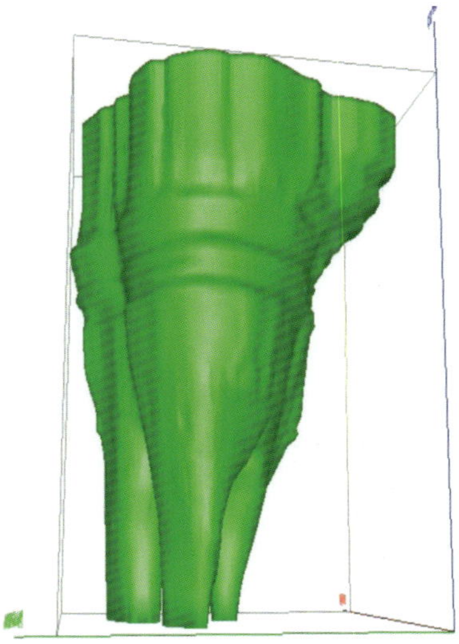

*Original Letlhakane Mine D/K1 pipe morphology with the 'three trouser legs' (Debswana)*

*Drilling dewatering boreholes at D/K1 Letlhakane Mine - Special corrosion resistant Roscoe Moss casing and screens are used to construct the boreholes (Debswana, 2010)*

# Mining

*Aerial view of Debswana Damtshaa Mine and Plant, B/K9 on left, B/K12 on right, 2006 (Debswana)*

*Damtshaa mine B/K12 pit (M. Brook, 2007)*

## DAMTSHAA MINE ("WATER FOR A TORTOISE")

Four small diamond pipes (B/K1, B/K9, B/K12 and B/K 15) were discovered between 1967 and 1972 in an area 20km east of the Orapa kimberlite pipe. These pipes were grouped to form the Damtshaa Mine, the youngest of the Debswana mines which began operation in 2003, and are managed along with Letlhakane Mine from the Orapa operation. The ore grade for B/K9 and B/K12, the kimberlites currently mined, is about 29 cpht and annual production is a maximum of about 330,000 carats. B/K9, at 11.5 hectares, is the larger of the two mined pipes and has a current area of excavation of about 21 hectares, and a depth below surface of 50m.

B/K 12 pipe area at surface measures 3.2 hectares and the current area of excavation is 7 hectares, with the pit bottom 63m below ground level. Combined with B/K9, about 2 million carats have been recovered to date. In 2008, a mining contract was awarded to a Joint Venture comprising Basil Read Mining and Bothakga Burrow Botswana. This mine, however, was temporarily closed at the end of 2008 due to the financial global crisis, but reopened in 2012. The life of the open pit ends in 2026 (B/K9).

*Production blast drill rig at Damtshaa Mine (Debswana, 2007)*

# Mining

*Debswana Jwaneng Mine Aquarium (Debswana, 2008)*

## JWANENG MINE ("WHERE A SMALL STONE IS FOUND")

The Jwaneng pipe was discovered in the Naledi River Valley ('Valley of the Stars'), southern Botswana, in 1972. At that time there were a mere 63 dwellers in the region called Jwana, the site of the present day Jwaneng Town. The mine is the world's largest diamond producer by value. The mine is located on the fringe of the Kalahari Desert and at start up recruited 1,345 employees. The township of Jwaneng was developed from scratch and today has a population of just over 18,000 and has all the amenities one would expect.

*De Beers Geologists H Roos, G Edwards, G Lamont, and G Curry discuss the planning of the new Jwaneng Mine (De Beers, 1977)*

*Balisi Bonyongo, General Manager of Debswana Jwaneng Mine from 2007 to present (Debswana, 2009)*

*New, lined water reservoir, Jwaneng Mine (Debswana, 2011)*

*Jwaneng D/K2 Open Pit (De Beers, 1982)*

*Debswana Jwaneng Mine Main Treatment Plant (De Beers, 1982)*

# Mining

*Jwaneng Mine (Debswana, 2011)*

*Ore conveyors, Jwaneng Mine (Debswana)*

*Official opening of Debswana Jwaneng Mine on 14th August, 1982, by Sir Ketumile Masire, President of the Republic of Botswana, 1980 - 1998, Mike Smith, first General Manager of Jwaneng Mine, Harry Oppenheimer, Chairman of De Beers and Sir Ketumile Masire. (De Beers, 1982)*

# Mining

*Debswana Jwaneng Mine. Main plant on left, pit centre, tailings dumps and slimes dams in background (Illustrative Options, courtesy Debswana, 2011)*

*A joint venture between Bauer and De Wet Drilling Botswana was formed in 2010 to design and manufacture drilling rigs in Botswana to service diamond and other mines. Shown above are the Rhino 2200 blast hole rig and the Buffalo 90/TBA 100T1 multipurpose deep drilling rig, launched in 2012. (De Wet Drilling, 2012)*

*Jwaneng Mine, Cut 8 (Debswana, 2011)*

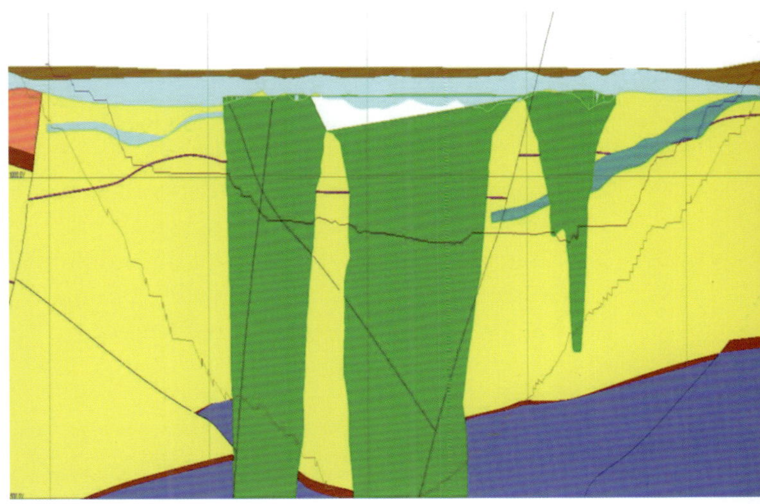

*Cross section of the three Jwaneng Kimberlite Pipes - Kimberlite (green), Kalahari sequence (top brown & light blue), Shales (yellow), Dolomite (lower blue)*

Jwaneng Mine is the main revenue driver in Debswana due to substantially higher dollar per carat obtained for its gem quality diamonds. Jwaneng Mine contributes about 60-70% of Debswana's total revenue.

# Mining

*Jwaneng Mine and Aquarium (Debswana, 2011)*

The Jwaneng mine Aquarium plant is the first of its kind in the diamond mining industry. It has the Completely Automated Recovery Plant (CARP) and the Fully Integrated Sort House (FISH). This project brought to an end the hand sorting of diamonds. High tech x-ray scanners used at this facility ensure quick and accurate diamond recovery and maximised profit. After sorting the diamonds from all Debswana operations, they are cleaned by a process of acidisation at Jwaneng.

*Jwaneng Mine (Debswana, 1982)*

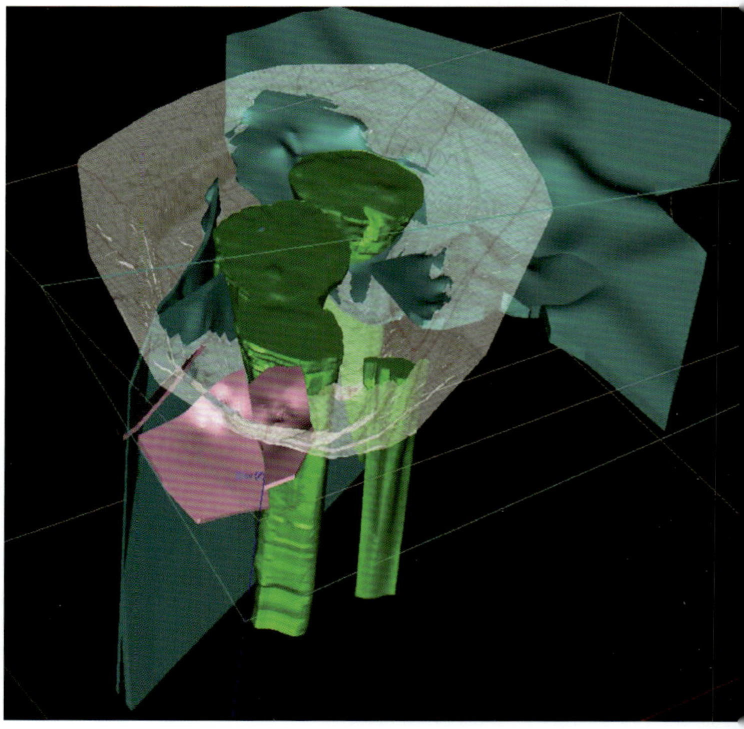

*3-D Geological Model of the three kimberlite pipes at Jwaneng Mine (Debswana)*

Currently Jwaneng is mining to a depth of 350 metres and is expected to reach 624 metres in 2017 as a result of Cut 8 mining. The resource consists of three separate volcanic pipes/vents namely north, south and centre pipes and a small kimberlite sill which erupted through Transvaal strata and the overlying Karoo sediments about 245 million years ago.

The area of the D/K2 pipe at surface is 54 hectares and the area of excavation at surface is 355 hectares. The central pipe is the richest and larger diamonds are thought to be associated with the southern pipe. The average diamond grade is 140 cpht. The pit currently measures 1.6km by 2.4km and the life of open pit is to 2029, excluding processing of the tailings dumps which have significant potential still for further diamond recovery. This extension of the pit life will be achieved by the current Cut 8 project completed in March 2012.

Production normally varies according to mining plans of approximately 12.5 to 15 million carats per year.

In 2010, Jwaneng produced 11.5 million saleable carats. In 2007, the maximum annual production of 20.3 million carats was achieved. In 2011, Jwaneng produced 10.6 million carats.

After the Cut 8 construction phase has been completed, ore production and treated ore volumes will remain at about ten-million tons a year. The Cut 8 project involves the relocation and rebuilding of conveyors, crushers and stockpiles and the demolition of existing structures that fall within the Cut 8 pit perimeter. Cut 8 will extend the open pit life of mine to the year 2026 and will recover approximately 100 million carats of mostly gem quality diamonds.

*Demolition of Jwaneng Mine In-Pit crusher to make way for Cut 8 (Debswana, 2011)*

# Mining

*Deep Diamond core drilling by De Wet Drilling, for the Jwaneng Resource Extension Project (M. Roberts, 2011)*

The mining of Cut 8, the largest out-sourced mining services contract ever awarded by Debswana, is being undertaken by MAJWE Mining Joint Venture, which consists of Basil Read Mining, Leighton International Limited and Bothakga Burrow Botswana. The five-year contract is for the Cut 8 Phase 2 services which include mine scheduling, drill and blast, truck and shovel waste removal and limited ore mining. MAJWE will move a staggering 156 million cubic metres of material over the full term of the contract, which commenced in June 2011.

According to Steve Axcell, then Debswana Deputy Managing Director, *"Studies are currently under way for the next phase of expansion at the Jwaneng mine, the Cut 9 expansion project, which will aim to further extend the life of the mine. These studies, which include deep diamond core drilling as part of the Jwaneng Resource Extension Project, are focused on optimising the cut design in terms of pit slope angles and will be implemented during initial Cut 9 expansion activities planned to start in 2016. This long-term expansion plan is in line with anticipated increased global diamond demand, which could grow steadily over the next few years."*

At the official launch of Cut 8 on the 10th December, 2010, His Excellency President Lt. Gen. Seretse Khama Ian Khama said. *"Jwaneng Mine is the flagship of Debswana's mining portfolio due to the substantially higher revenue obtained for its gems.*

*I am reliably informed that the Cut 8 Project will play a significant role in maintaining Jwaneng Mine's status as the richest diamond mine by value in the world. The mining of the 8th cut or Jwaneng Cut 8 as it is commonly known, will cost P24 billion over a period of 16 years from the year 2010, inclusive of all project capital, mining fleet capital, sustaining capital and all working costs. This will not only ensure sustainability for Jwaneng Mine but also the resultant creation of jobs as well as revenue for Debswana's shareholders. It also showcases Botswana's economic capacity and potential to undertake a project of this size during recessionary times".*

Jwaneng Mine is also surrounded by a game park, "Jwana Park", which is managed by Debswana.

Debswana employs over 4,500 people and is Botswana's largest private sector employer. Out of every five pula derived from diamond revenues, the Botswana Government receives four pula through dividends, royalties and taxes.

*Geologists on the Jwaneng Resource Extension Project discuss the results of core drilling (Debswana, 2008)*

*Rhino and Zebra at Debswana Jwana Park (Illustrative options, courtesy Debswana, 2008)*

*Giraffe (G. Beale, 2012)*

# Mining

*Early Mining at Lerala Mine (DiamonEX, 2008)*

*Lerala Plant (M. Brennan, Mantle Diamonds, 2012)*

*Lerala Mine Ore Conveyors (DiamonEX, 2008)*

## LERALA MINE

The five kimberlites which make up this mine in the Central District were discovered by De Beers in 1991 but it was only in 2008 that a mine was established. The 330,000 carat per year diamond mine, originally owned by DiamondEx Limited, was fully commissioned in June 2008 at a cost of US$ 24 million. The mine comprises five kimberlite pipes with a combined size of six hectares. The pipes are exposed at the surface and the indicated resource is 13.5 million tones of kimberlite at a grade of about 25.5 cpht. Life of mine is put at 10 years, but could possibly go deeper after that.

The mine was put on a care and maintenance basis during the 2008 global financial crisis and has yet to re-commence mining. In 2011, Lerala Mine was bought by Mantle Diamonds UK for US$3,250,000 and the latter have started preparations to re-open the mine in 2012.

The government got involved in this project during the global economic crisis when it guaranteed DiamonEx Botswana a 10 million pula (about 1.52 million USD) bridging loan facility from commercial banks and engaged some of the creditors in a bid to resolve the company's financial problems to preserve the future of the mine and attract new investors.

*Lerala Mine Processing Plant (DiamonEX, 2008)*

# Mining

*B/K11 Secondary crusher being commissioned (Firestone Diamonds, 2011)*

## B/K11 MINE

B/K11 is located approximately seven kilometres north-west and 20 kilometres south-east of Debswana's Letlhakane and Orapa mines respectively, and is within five kilometres of Karowe Mine at A/K6 which opened in 2012. The surface area of B/K11 is estimated to be eight hectares, and overburden is shallow at less than 20 metres. The pipe was originally discovered by De Beers in 1970, but was deemed to be uneconomic at the time.

*Production blast drilling at B/K11 pit (Firestone Diamonds, 2011)*

Mining of this single kimberlite by Firestone Diamonds commenced in July 2010. Full production of about 10,000 carats per month was reached in 2011. Its diamonds are valued at $175 per carat and with an 11 million ton resource at 8.5cpht, the value of the resource is estimated at about $150 Million. The first auction sales of B/K11 diamonds were held in Gaborone in 2011 and grossed $1.35Million for the first quarter. The second quarter's diamond auction attracted 31 companies and grossed $2 Million. The largest diamond recovered to date is a high quality 13.74 carat stone. The mine is expected to have a life in excess of 10 years.

Firestone Diamonds are also evaluating another new potential mine development, B/K16, about 22km north - east of B/K11 and 12km north of Letlhakane Mine. This pipe is estimated to be at least 3.5 hectares in size and has been proved to contain 17 million tonnes of kimberlite to a depth of about 200m. Firestone also controls the rights to 21 other satellite kimberlites in the Orapa area, of which eight have been proven to be diamondiferous.

*Processing Plant at B/K11 (Firestone Diamonds, 2011)*

# Mining

A/K6 Open Pit Karowe Mine (Boteti Mining, 2012)

Large diameter ore evaluation drilling by De Wet Drilling (De Wet Drilling, 2005)

A/K6 processing thickener (Boteti Mining, 2011)

## A/K6 (KAROWE MINE)

The Boteti A/K6 kimberlite is an advanced, high value diamond development project located in the Orapa/Letlhakane kimberlite district of Botswana. Lucara holds a 100% interest in the A/K6 project.

The Boteti A/K6 kimberlite consists of three lobes, South, Centre and North, of which the South Lobe makes up approximately 75% of the kimberlites' resource potential. The pipe has an area of 4.2 hectares at the surface which expands to seven hectares at a depth of 120 meters. From an Indicated Resource of 51 million tonnes containing 8.2 million carats of diamonds, the mine design delineates Probable Reserves of 36.2 million tonnes of ore, containing 6.3 million carats of diamonds in an open pit to a depth of 324 metres. With a grade of about 16 cpht, the reserves will be mined over a period of 11 years.

Construction of the A/K6 diamond mine required a capital investment of approximately US$130 million, which includes the process plant and all mine site and off-site infrastructure. Instead of using conventional scrubbers to disaggregate the kimberlite ore in order to liberate the diamonds, this mine uses an AG or autogenous mill (self generating mill). The AG mill acts as a high intensity ball which produces acceleration fields 40 to 50 times stronger than gravity, producing rapid and intense breakage of kimberlite ore through high intensity tumbling. Production for 2012 is estimated at 400,000 carats which will be sold in Gaborone through a tender system. In 2011, the mine was given the name of Karowe Mine, a Sesarwa name meaning a precious stone.

*A/K6 AG Mill (Boteti Mining, 2011)*

*A/K6 Primary Crusher (Boteti Mining, 2011)*

# Mining

*Pre-Mine, Ghaghoo Mine (Gem Diamonds 2010)*

*Setting up an Open Faced Cutting Shield at Ghaghoo Mine (Gem Diamonds, 2011)*

*First 30m of the incline at Ghaghoo Mine (Gem Diamonds, 2011). The cutting shield progresses at a rate of about 3.6m per day.*

## GHAGHOO MINE (FORMERLY GOPE MINE)

Gope is named after the region of the Central Kalahari Game Reserve (CKGR) where the Gope 25 kimberlite pipe is located 45km within the eastern border of the CKGR. Gem Diamonds acquired Gope Exploration from De Beers and Xstrata in May 2007 for US$34 million. In 2011, the name of the mine was changed from Gope – "Nowhere" to Ghaghoo – the Sesarwa name for the locally abundant tree – Acacia Leuderitzii – False Umbrella Thorn.

A mining licence was approved in 2011 and production is expected to commence in 2013 with the life of mine currently estimated to be in excess of 30 years. Gope's total resource is estimated at 20.5 million carats with a total average in situ value of US$3.3 billion. The average resource diamond price has been increased to US$162 per carat at a grade of about 22 cpht. Two blue diamonds were also recovered from the historical diamond samples during the course of valuation work.

*General layout of Ghaghoo Mine (Gem Diamonds, 2011)*

In 2011, there was a change in mine plan from open pit to underground when a box cut and 8% decline construction commenced which would expose the top of the kimberlite ore body at a depth of 150m below ground level, covered by about 80m of Kalahari sand and 70m of basalt.

The (ZAR) 67-million contract for the development of the sand tunnel was awarded to contract mining firm, Redpath South Africa. The contractor is making use of cost-effective conveyor belts for the removal of the sand, as opposed to articulated dump trucks. Although the initial set-up costs of the conveyor belts were high, the long-term running costs are significantly-lower than using trucks. The box cut will lead into a concrete-lined segmented sand tunnel, 403m in pure sand and 121m in transition from sand to basalt, which acts as a cylindrical shield to protect workers and equipment located underground.

## OTHER MINERALS

Nickel, copper, cobalt, soda ash, salt, coal (currently, the most important resource after diamonds, with tremendous prospects for numerous new coal mines to be developed), semi-precious stones, crushed aggregate, clay and sand are also mined in Botswana, but with the revenue generated accounting for more than 60% of government's coffers, diamonds are by far the most important mineral sustaining Botswana's growing economy. Other potentially valuable minerals include iron, zinc, uranium, lead, cobalt, antimony, chromite, feldspar, fluorite, graphite, gypsum, limestone, managanse, platinum,soapstone and talc.

Having now recovered the ore from the mines, the next stage in the diamond pipeline is to process the kimberlite in order to liberate the diamonds.

MORRIS
CRANES

# 6 Processing and Recovery

Interestingly, the methods of processing the kimberlite ore and recovering diamonds are based on processes and technology which take advantage of some of the unique properties of the diamond mineral itself.

The first, being the hardest and most robust natural material on earth, means that the mineral is resilient and can survive the first rigorous process of primary, secondary and re-crushing the kimberlite ore into smaller fragments.

*Orapa Mine Primary Crusher (Debswana, 2002)*

# Processing and Recovery

The second property of the diamond which metallurgists have used in the recovery of the stones, is that of its heavier specific gravity - at 3.52 g/cm$^3$, it is 3½ times heavier than water and also heavier than most other minerals and as such, diamonds sink to the bottom of the chamber called a dense media separator, where they can be concentrated before undergoing X-ray processing.

This brings us to a third property of the diamond which is the fact that it is transparent to X-rays and luminesce or emit light when exposed to X-ray radiation.

When the above dense media separator concentrate is passed through beams of X-rays, diamonds show up as white objects.

The X-ray technology, upon detecting the diamonds, then jets a forceful spray of air which throws the diamonds, and some surrounding concentrate, into containers which are then sorted further for the recovery of the diamonds.

In some smaller diamond processing plants, and diamond sampling plants, a "grease table" belt is utilised which exploits another interesting characteristic of the diamond -the fact that it cannot be wetted - it repels water, just like rainwater off a duck's back, and sticks to the layer of grease coated on the conveyer belt where it is recovered by melting the collected grease once all the ore has been processed.

Lets take a closer look at the processing technology currently used to recover diamonds from the host rock ore, kimberlite.

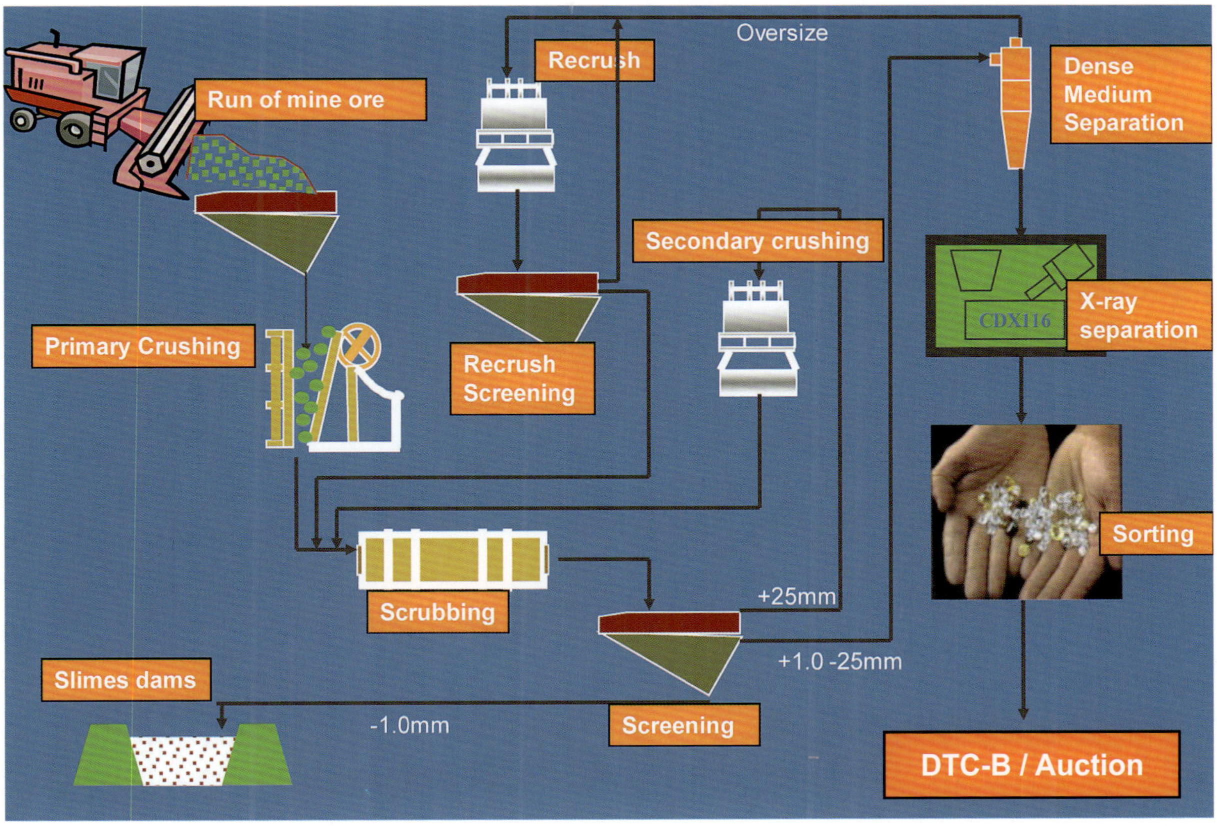


*A typical Diamond processing plant flow chart (Debswana)*

# Processing and Recovery

Debswana Orapa Mine Primary Ore Crusher (Debswana)

Although there are various diamond recovery technologies being utilized worldwide today, the two principal types are Rotary Pan plants and Dense Media Separation (DMS). Both these methods are in actual fact used for overall bulk material reduction and require yet another process before finally recovering the diamonds. These systems are based on the property of diamonds having a much higher specific gravity density, as discussed above, compared to most of the other minerals which are found in situ with diamonds.

The first stage of kimberlite processing is to crush and screen the mined ore as required in order to reduce its size. The first, or primary crusher, which may be located either in - pit or at the surface of a mine, reduces the rock size from <1m to >150mm. Some boulders several meters in size may be crushed in this process. The reduction of the ore size by crushing therefore facilitates material handling and the crushed rock is then transported by rubber conveyor belts for treatment through subsequent processes, including a secondary crusher which reduces the kimberlite rock size from <150mm to >25mm. This crushing helps to liberate (free) diamonds from the rock matrix.

Damtshaa Mine Ore Conveyors (Debswana, 2007)

*Jwanang Cut 8 replacement ore conveyor (Debswana, 2011)*

A third stage of crushing, termed re-crushing, may also be used e.g. Jwaneng Recrush (inter-particle crusher) plant (commissioned in 1991) which then reduces the rock particle size from <25mm to >8mm. This plant can also be used to re-crush the coarse DMS tailings (original reject material from the first crusher pass concentrate through the DMS plant). This re-crushing liberates finer locked diamonds.

Depending on the processing method being employed, material under a certain chosen size fraction may go through a scrubbing system to clean the ore prior to it being fed into the processing plant. Scrubbing basically comprises a tumbling action in a washing drum which helps disperse associated clay / agglomerated material. This scrubbing action produces a clean ore feed for efficient separation in the DMS process and also liberates entrapped diamonds from briquettes.

The next stage of processing involves screening which separates solid particles on the basis of size. Right size fractions are then produced for subsequent processing.

*Jwaneng Mine Main Treatment Plant and Aquarium (Debswana, 2011)*

# Processing and Recovery

*Jwaneng mine re-crushing plant. Tumbling of briquettes of ore with water in the Scrubber disagglomerates the ore and liberates the diamonds (Debswana, 2007)*

*Ore screening at Jwaneng Mine (Debswana, 2007)*

The final stage of diamond recovery, once the above preparatory stages have successfully been completed, is using either Rotary Pan plants or Dense Media Separation (DMS) plants.

## ROTARY PAN PLANTS

In a Rotary Pan plant, crushed ore, when mining kimberlite, or alluvial gravel and soil is mixed with water to create a liquid slurry called "puddle" which has a density in the 1.3 to 1.5 g/cm$^3$ range. The mix is stirred in the pan by angled rotating "teeth". The heavier minerals, or "concentrate", settle to the bottom and are pushed towards an extraction point, while lighter waste remains are suspended and overflow out of the centre of the pan as a separate stream of material which is discarded. The concentrate, representing just a small percentage of the original kimberlite ore or alluvial gravels, is drawn off for final recovery of the diamonds by a sorting process.

# Processing and Recovery

*Debswana Jwaneng Mine Main plant, seen from first benches of Cut 8 (Debswana, 2011)*

## DENSE MEDIA SEPARATION (DMS) PLANTS

In a Dense Media Separation (DMS) plant, powdered ferrosilicon (an alloy of iron and silicone) is suspended in water to form a fluid very close to the same density of diamond (3.52 $g/cm^3$), to which the diamond bearing material is added to begin the separation process of the heavier minerals from the lighter material. Additional separation of the denser material occurs by centrifuge in"cyclones" that swirl the mixture at low and high speeds, forcing the diamonds and other dense minerals to the walls and then out the bottom of the cyclone where it is collected.

Wastewater rises at the center of the cyclones and is sucked out and screened to remove waste particles. The DMS process results in a concentrate that generally weighs less than one percent of the original material fed into the plant at the beginning of the process, and which includes a very high percentage of the diamonds originally present. The DMS tailings or discarded material may then be fed through the same process a second time to ensure complete diamond recovery is attained. The final product contains mostly diamonds, with minor quantities of other material. It is fed into lockable canisters and then transported to a diamond sorting facility, such as the new Diamond Trading Company Botswana's state of the art sorting and valuation center in Gaborone.

Inside Jwaneng Mine Aquarium (Debswana, 2008)

Both recovery methods have their advantages and disadvantages. The capital required for an equivalent (high) throughput DMS plant is in the order of ten times higher than the capital required for a Rotary Pan plant, but can typically offer better recoveries. Water usage and operating costs for an equivalent DMS plant are also typically much higher than that of a Rotary Pan plant. Fortunately, all of Botswana's diamond mines have sufficient water supply, however, being a semi-arid country with scarce water resources, all plants must be designed to be able to recover and re-use as much water as possible. This is done by large circular "thickeners" which thicken the process waste residue so that as little water as possible is then used to pipe the residue to the "slimes" or collection dams.

# Processing and Recovery

*Jwaneng Mine Paste Thickener (Debswana, 2007)*

*Orapa Paste thickeners (Debswana, 2006)*

*Orapa Mine Slimes Dam (M. Brook, 2008)*

At the dams, cyclones are used to separate water from residue which is then sprayed to build up the actual walls of the dam. The clear water is then centrifuged to the centre of the dam where it decanters so that the recovered water can then be pumped back to be re-used in the processing plant.

# Processing and Recovery

*Blast prepared at Orapa Mine (Debswana, 2006)*

Rotary Pan plants are most often employed when mining alluvial deposits and DMS plants for kimberlite deposits. A primary kimberlite deposit generally has a much longer life-of-mine than an alluvial deposit and due to the nature of the deposit, allows for the development of stationary infrastructure. Thus the higher capital costs associated with a DMS plant are more easily justified.

*One of the X-ray diamond recovery machines in the CARP*

*Diamond recovery by hand, Red Area, Orapa Mine (M. Brook, 1984)*

A special mention needs to be made of the state of the art recovery and sorting building at Jwaneng, known as the "AQUARIUM", which was commissioned in the year 2000.

This tall set of four building structures contains a Completely Automated Recovery Plant (CARP) which is completely "hands-off" and therefore maximises security while optimising diamond recovery through the use of the latest X-ray machine technologies.

*Depositing Tailings on the Orapa Mine tailings dump (Debswana, 2009)*

*Ferrosilicon recovery from the magnetic separator drum. The Ferrosilicon is recovered from the gravels passed through the dense media separation process and is re-used. (Debswana)*

CARP produces concentrate with > 50% diamond by weight. The CARP was designed to treat concentrate from the Jwaneng Main Treatment Plant (MTP), and the Jwaneng Recrush Plant, and tailings from the Jwaneng old recovery tailings dump.

*Orapa CARP building (Debswana, 2002)*

# Processing and Recovery

*Debswana Jwaneng Mine Aquarium (Debswana, 2008)*

Within the same "AQUARIUM" buildings, diamond sorting, cleaning, packing and weighing takes place in a Fully Integrated Sort House (FISH) which is also completely "Hands – Off" and maximises diamond security while optimising diamond sorting through the use of laser technology. FISH is more efficient than hand sorting for certain product categories. Laser sorting is based on measuring in-elastically scattered light from molecules and this is referred to as Raman spectroscopy. The Raman scattered light occurs at wavelengths that are shifted from the incident light by the energies of molecular vibrations. (du Plessis and M. Sewawa, 2003).

Single Particle Sorting (SPS) is undertaken using laser technology. The feed is concentrate from CARP (>50% diamond by weight) and particles are presented one at a time and detected using a laser beam. The resultant SPS product is a concentrate in excess of 95% diamond by weight. The FISH was designed to treat all Jwaneng sources, as well as diamond concentrate from Orapa CARP, consisting of concentrates from Orapa No. 1 processing plant, Orapa No. 2 processing plant, Letlhakane Mine and Damtshaa Mine. Orapa sources are delivered on a weekly basis, while Jwaneng sources are treated as they become available.

*Single particle sorter in the FISH - single particles are assessed individually to identify diamonds (Debswana)*

# Processing and Recovery

*Damtshaa Mine Treatment Plant (M. Brook, 2007)*

*Jwaneng Aquarium Plant (Debswana, 2010)*

Acid cleaning of the final SPS product is also done at FISH. Acidation using Nitric and Hydroflouric Acid dissolves silicates or minerals attached to the diamond surface; silicates dissolve in acid but diamonds do not dissolve. Further diamond cleaning is then done at the Jwaneng Central Acidising Centre (CAC). The CAC is used to clean all of Debswana's diamond production.

The final CAC product (>99% diamond by weight) is then sent to the Diamond Trading Company Botswana in Gaborone for final sorting and valuation. Other improvements achieved by the Aquarium are improved information turnaround time and reduced operating costs.

Having successfully liberated the diamonds, the next stage in the diamond pipeline is to sort and value all the diamonds recovered.

# 7 Sorting and Valuation

Dealing in diamonds in Botswana falls under statutory instrument 24 of 2004 of the Precious and Semi- Precious Stones (Protection) Act (Cap 66:03) of the Mines and Mineral Act, Chapter 66:-1, 1999 while Diamond Cutting falls under Cap 66:04.

In the same year that Botswana's first diamond mine at Orapa was established in 1971, the Government of Botswana formed the Botswana Diamond Valuing Company (BDVC), in Gaborone, the capital city of Botswana. At the time, second only to the Diamond Trading Company (DTC) in London, BDVC was the second largest rough diamond sorting and valuing operation in the world. BDVC was eventually registered as a company in 1972 with De Beers and Botswana Development Corporation as partners, to carry out the preliminary sorting of diamonds from Orapa Mine, which was the only operational mine at the time.

# Sorting and Valuation

*BDVC Headquarters - Orapa House, Gaborone (BDVC, 2002)*

In 1977, BDVC became a wholly owned subsidiary of Debswana. Up until it was replaced by the Diamond Trading Company Botswana (DTCB) in 2006, BDVC had a work force of more than 500 people.

BDVC sorted and valued Debswana's entire diamond production of diamonds from the Jwaneng, Orapa, Letlhakane and Damtshaa mines, and associated various prospecting diamond deposits. In 1982, Botswana Diamond Valuing Company Sorting building, nicknamed "Harry Oppenheimer House" but formally known as Orapa House was opened in Gaborone and was the tallest and most sophisticated building in Botswana at the time. Window panes on the ten floors were angled to allow just the right quality of natural light for assisting in diamond sorting and valuation. After sorting and valuation, diamonds were sold to the De Beers Central Selling (CSO) Organisation in London, which has now been replaced by Diamond Trading Company International, London.

*Opening of BDVC Orapa House, 1982 by Dr. G. Chiepe and Mr. H. Oppenheimer (De Beers, 1982)*

*Sorting diamonds at BDVC, Gaborone (BDVC, 2002)*

# Sorting and Valuation

*World's largest diamond and valuing sorthouse - DTC Botswana (DTCB, 2008)*

Diamond Trading Company Botswana (DTCB), a 50:50 joint venture partnership between the Botswana Government and De Beers, replaced BDVC in 2006. Before the DTC purchases diamonds from Debswana, a process of agreeing the value of a diamond takes place. Both the DTCI and the Botswana Government diamond valuers (GDVs) are represented by diamond sorting experts to ensure fairness. Both the DTCI and GDVs must agree on how the diamonds have been categorised against a master sample. If either party believes a diamond belongs in a different category, they must justify this by reference to this sample.

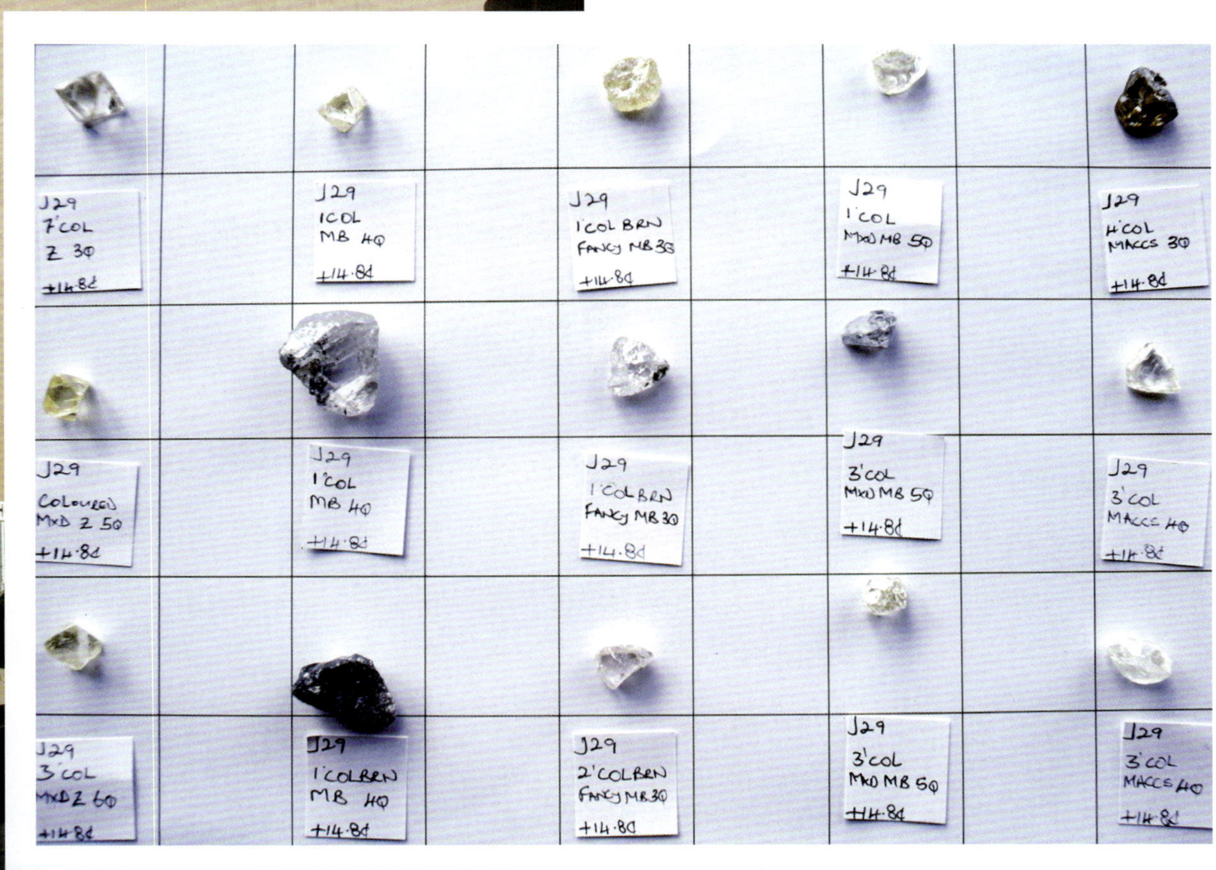

*A diamond sorting chart for large diamonds at BDVC (BDVC, 2002)*

# Sorting and Valuation

*Dr. Akolang Tombale and Charles Tibone, Botswana Government, and Nicky Oppenheimer, De Beers, signing the agreement to form DTCB on 23rd May 2006. (De Beers)*

Each of the categories corresponds to a certain price in the DTCI diamond price book. Once there is agreement on categorisation, the purchase process is complete.

On May 23rd, 2006, the Botswana Government and De Beers signed agreements to establish the Diamond Trading Company Botswana (DTCB), a 50:50 joint venture between the Botswana Government and De Beers and which replaced the BDVC in its entirety. In February 2006, the DTCB Diamond Academy was launched and on March 18th, 2008, the new DTCB building opened on the International Airport road adjacent to the Debswana Corporate Centre formerly known as the Technical Support Centre and became the largest rough diamond sorting and valuation facility in the world funded by De Beers at a cost of US$83 million.

It was officially opened in 2008 by the Botswana Minister of Minerals, Energy and Water Resources, Hon. Dr. Ponatshego Kedikilwe and Nicky Oppenheimer, Chairman of De Beers.

*DTCB building under construction (DTCB, Courtesy, Peter McMorran, 2007)*

*DTCB Sorting Building, Gaborone (DTC, 2008)*

*The wife of former President, Masire, Dr. Chiepe and Mike Brook visiting DTCB in 2009 (DTCB, 2009)*

# Sorting and Valuation

*Joesph Kabila, President of the Democratic Republic of Congo, visiting DTCB in 2011, (Illustrative options, courtesy DTCB, 2011)*

*Diamond Trading Company Botswana Acting Chief Executive, Paul Rowley, gives South African President, Jacob Zuma and First Lady, Sizakele MaKhumalo Zuma a tour of the largest diamond sorting and valuation building in the World (DTCB, 2012)*

*Xi Jimping, Chinese Vice President visiting DTCB in 2011, on his left, Minister of Minerals, Energy and Water Resources, Ponatshego Kedikilwe, and right, Brian McDonald, MD DTCB (Illustrative options, courtesy DTCB, 2011)*

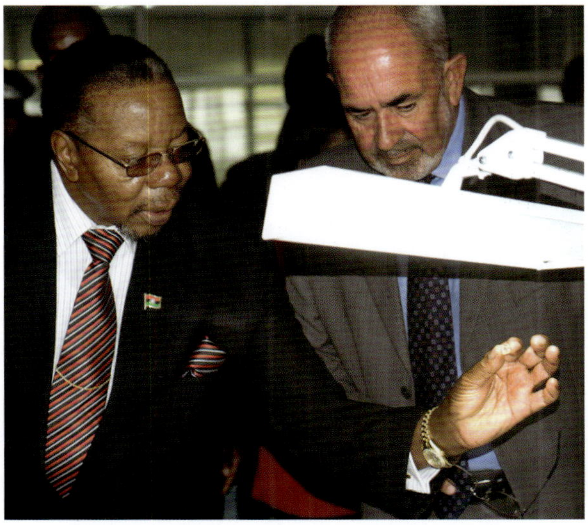

*The late HE Prof. Bingu wa Mutharika, President of the Republic of Malawi visiting DTCB in 2011, and right, Brian McDonald, MD DTCB (Illustrative options, courtesy DTCB, 2011)*

At the grand opening, Nicky Oppenheimer, Chairman of De Beers said, *"What a fantastic evening – what a fantastic venue. There can certainly be no greater monument to the on-going and enduring partnership between the people of Botswana and De Beers than this building that will be officially opened this evening"*. He further added *"What we are embarking on is nothing less than one of the largest transfers of skills and commercial activity to Africa ever seen. The diamond industry's center of gravity is shifting and tonight we see it shifting here"*.

The building is a world-class, high security diamond sorting house, having a gross superstructure floor area of approximately 21000m$^2$, together with a basement area of approximately 10000m$^2$, to accommodate and facilitate the functions of the DTCB.

Hailed also by (former) President Festus Mogae as a *"dream come true"* for the country, the specially designed giant green glass and brilliant chrome building, equivalent in size to six rugby fields, dominates the Gaborone skyline as one drives from Sir Serestse Khama International Airport. The building is fitted with the state of the art and most advanced diamond sorting technology in the world and has 39 diamond sorting machines, able to automatically distinguish between diamond gemstones in 29 different colours – more than is possible with the naked eye. Unlike its predecessor, the BDVC, which only sorted and valued diamonds, DTCB also performs local sales and marketing for rough diamonds.

The specialised sorting equipment was designed, manufactured and installed by engineers from the De Beers Research & Development division and the equipment is valued at about US$10M. The building has capacity for sorting and valuing 45 million carats per year and is able to accommodate up to six hundred employees.

# Sorting and Valuation

*Gem quality stones from DTC (De Beers, 2008)*

Former President, Festus Mogae, speaking at the official launch of DTCB and the new DTCB building said *"For the citizens of Botswana, it will bring increased employment opportunities. It will provide government with a significant revenue stream to finance further development. It will also raise our global profile"*.

*A heart of diamonds at DTCB (DTCB, 2011)*

Through its Diamond Academy in Botswana, the DTC ensures skills transfer, training and development of local employees. DTC Botswana is the primary vehicle for creating a sustainable and profitable downstream industry in Botswana that will:

- Provide maximum long term value from diamonds to the country through world class sorting, valuing, selling and marketing practices in Botswana
- Develop new employment opportunities in Botswana
- Ensure that new skills and capabilities are created amongst Botswana's citizens
- Be a catalyst for other businesses to invest in Botswana

# Sorting and Valuation

*High quality gems at DTC (De Beers, 2007)*

Today diamonds from the Debswana mines arrive in Gaborone for sorting, however, by the end of 2013, when DTCI London's sales and aggregation function re-locates from London to Gaborone, it will additionally receive diamonds from South Africa, Canada, Namibia and any other potential mines to be opened by De Beers or from other parts of the world. Diamonds are sorted as they arrive from the mines into different categories depending on size, shape, quality and colour, otherwise commonly known as the 4 C's – carat, cut, clarity and colour, a grading system which was invented by the Gemological Institute of America back in 1953.

*DTC Gems (DTC, 2007)*

# Sorting and Valuation

*Yellow and Brown Gems at DTC (De Beers, 2007)*

Today, DTCB currently sorts diamonds into around 12,000 categories. The sorting process ensures the correct valuation of all production. It also ensures that a consistent supply of diamonds can be delivered to the DTCB's customers (called Sightholders), who are among the world's leading diamantaires. In December 2011, the number of DTCB Sightholders was increased from 16 to 21, reflecting the increased interest and confidence that the world diamond business has in Botswana. Their rough diamond supply contracts were signed for a period of three years to 2014.

Diamonds are separated into two different types; namely Type I and Type IIa. Type I diamonds contain nitrogen while Type IIa diamonds are nitrogen free and are usually white, brown or pink and very rarely red or green. Type IIa and Type IIb (blue colour) are very rare and generally thought to comprise less than 1% of all diamonds. Gem quality diamonds are based on the combination of the 4C's characteristics described above.

In terms of colour, a range of white, moving to lighter yellows and browns are regarded as gem colours, with the whiter colours being the highest value colours. Deeper colours of yellow and brown, as well as unusual colours such as orange, violet, purple, green, pink, blue and red are referred to as fancy colours.

*A multi-coloured selection of shapes and sizes from DTC (De Beers, 2007)*

Size does matter, especially with large white D, E F & G colours. However, when dealing with rare fancy colours such as reds, blues, oranges and greens, even small stones are highly prized gems.

The standard reference for size is 1.00 carat = 200 milligram = 0.20 gram.

Originally gem quality rough diamonds were those diamonds without cracks, twins or large inclusions, the equivalent of the polished grade SI3 or better.

In today's terminology, the better near-gem qualities are included as gem quality, as it is now viewed that any diamond which can provide a polishable gem be included, regardless of the fact that more splits (sawing, cleaving or lasering) are required to achieve polishable pieces.

The most desired rough gem shapes are the following:

- crystals (Sharp edged octahedrons),
- octahedrons with some rounding to the edges and points
- dodecahedrons
- triangular maccles
- rounded dodecahedral maccles
- shapes that have high yielding forms either for rounds or fancy cuts

Less desirable gem shapes are cubes, irregular low yielding shapes and twinned or aggregated forms.

Therefore, the most desired rough gem diamonds would be those having good colour, shape and internal clarity, without ignoring the size factor.

# Sorting and Valuation

*Large sorted stones at BDVC, Gaborone (BDVC, 2002)*

The first step in the sorting process is to consider the size of incoming diamonds from the so called "Run of mine" production which arrives in specially sealed metal canisters. There are three basic forms of diamonds namely Octahedrons, Cubes and Dodecahedrons.

Diamonds weighing more than 10.80 carats are classified as Large Stones or Special Stones. Each stone is assessed by hand by experts to determine its value. A trained eye and expert judgement is used in conjunction with the latest diamond imaging technology to correctly identify and value the diamond based on the expected polished content of these large stones.

# Sorting and Valuation

*Checking diamonds at BDVC (EDVC, 2002)*

*An Africa of diamonds (DTCB, 2011)*

The next size category is for diamonds weighing between 0.66 carats and 10.79 carats which are divided into a number of weight categories using sorting technology designed and manufactured by the Diamond Trading Company (DTC).

Finally, diamonds weighing less than 0.66 carats are divided into a further six categories using sieve plates.

# Sorting and Valuation

*Sorted diamond samples at BDVC (BDVC, 2002)*

*Small white gems at BDVC (BDVC, 2002)*

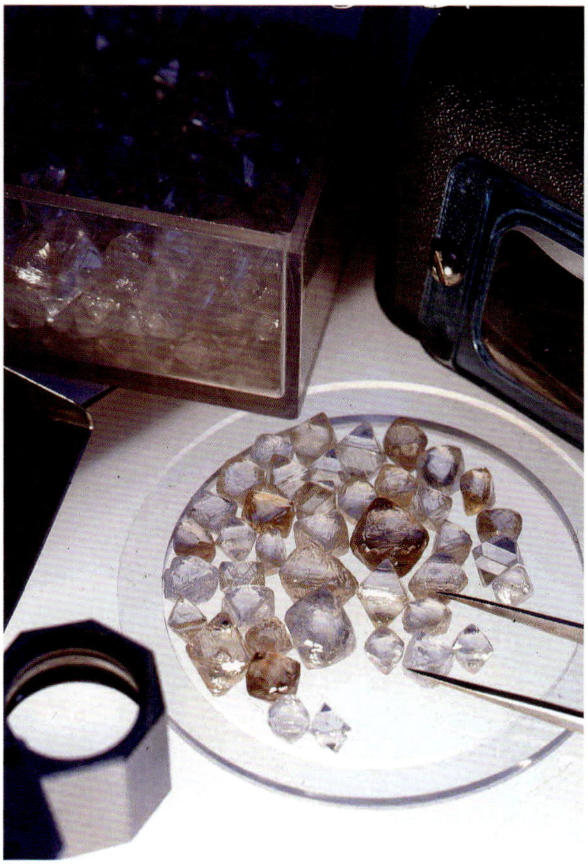

*Fancy colours and a diamond sorters toolkit, DTC (De Beers, 2007)*

# Sorting and Valuation

*President Masire and Julian Ogilvie Thompson, Chairman of Anglo American Corporation, at BDVC Orapa House with Botswana's largest single diamond discovered to date. The 446-carat gem was found at Jwaneng Mine in 1993. (De Beers)*

Unlike South Africa and Lesotho, Botswana is not renowned for the recovery of very large diamonds. Some very large diamonds have been recovered from mining but very few have remained intact. These larger diamonds have all been mined from Jwaneng Mine. In 1994 and 2011, several pieces of the same diamond were re-constructed to produce +600 carat stones. The largest broken diamond mined was in the year 2000 and comprised 13 pieces, the largest of which was +100 carats, which when reconstituted, formed a single diamond of 650 carats weight. The largest single, in-tact diamond also mined to date in Botswana also came from Jwaneng, in 1993, and weighed 446 carats.

The largest rough diamond ever discovered in the world is that of the Cullinan Diamond, which weighed 3,106 carats, found at Premier Mine, South Africa in 1905. From it, nine large stones were cut and polished, the largest being Cullinan I, weighing 530 carats. This stone was placed by King Edward in the Sovereign's Royal Sceptre (crown) as part of the Crown Jewels of the British Monarchy and is now on display in the basement of the Tower of London.

The second largest rough diamond discovered, from the Jagersfontein Mine, South Africa in 1893, weighed 995 carats and is called the Excelsior diamond. The largest ever cut and polished stone to date is the Golden Jubilee diamond which weighs 545 carats and was cut from a 755 carat rough stone found at Premier Mine, South Africa, in the year 1985.

# Sorting and Valuation

*Small white gems at DTCB (DTCB, 2008)*

The second step in the sorting process is termed "Model" which refers to the shape of the stone. The shape of the rough diamond often dictates the final shape and size of the polished stone. Rough diamonds fall into two main model categories, "Sawable" and "Makeable". Sawable stones are often octahedral or dodecahedral in shape. The rough diamond is sawn prior to being polished. This will produce two polished stones.

Makeable diamonds tend to be broken octahedrons or dodecahedrons or flatter whole models. They usually produce one polished stone, but may require more work than Sawable models, however less time overall is used in their manufacture as no sawing is required. Small and more complicated, lower value diamonds may need to be split into small useable pieces using traditional cleaving or laser technology.

*A collection of rough stones exhibiting a 'greasy' appearance (DTC)*

# Sorting and Valuation

*Sorted and beautifully shaped DTC gems (De Beers)*

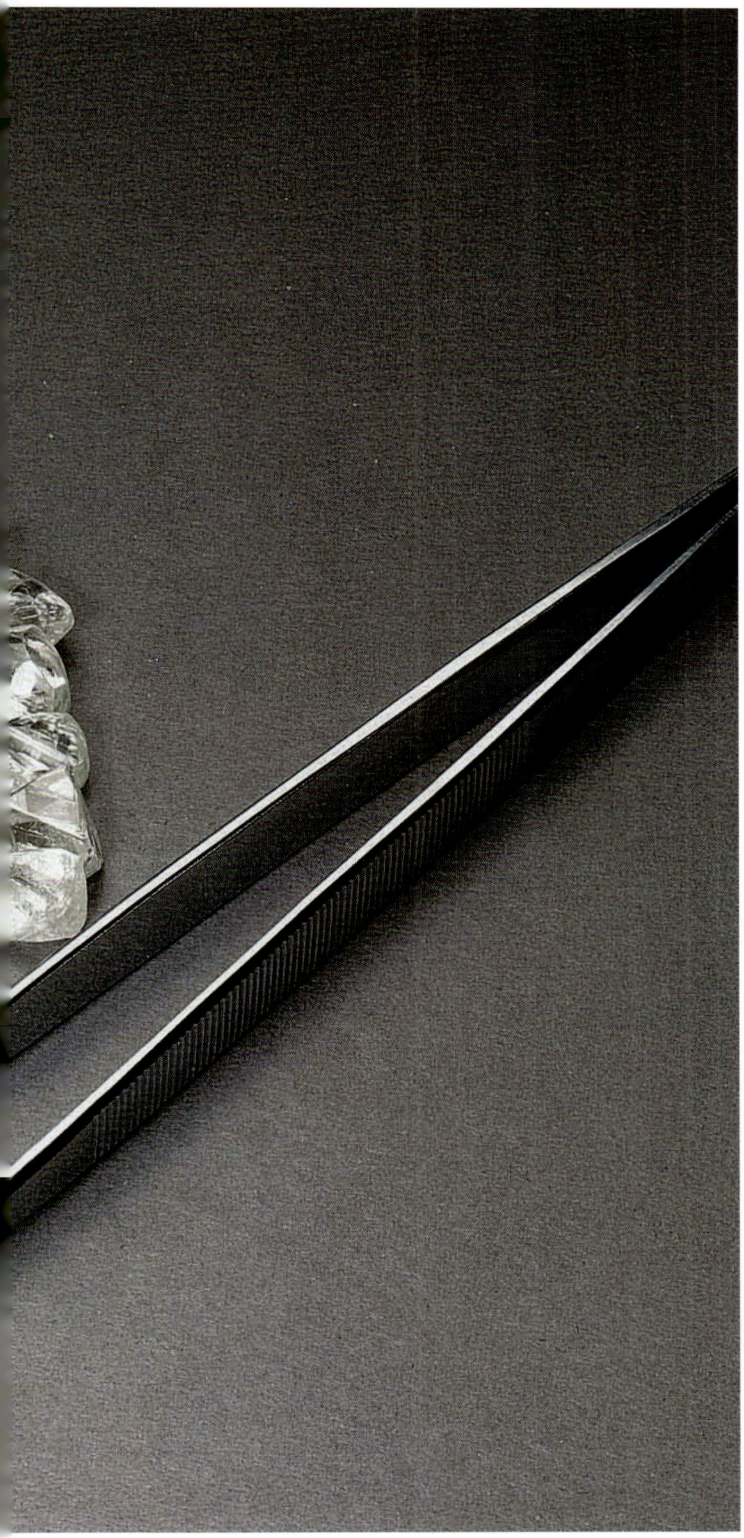

The third step in the sorting and valuation process is to decide on the quality of the diamond. This is a measure of the number, size and type of impurities, inclusions or cracks found within the stone. This significantly affects the final value of the diamond and may impact on the way it is split up into smaller stones prior to polishing. Inclusions also impact the way in which light travels through the stone, which could reduce the "fire" and brilliance of the polished diamond. An inclusion free, clean stone holds greater value because of its rarity.

There are two broad categories in quality sorting - "gem" and "near gem".

When grading gem quality rough diamonds, the number, location, variability and final impact on polishing of the inclusions is carefully assessed.

When grading near gem quality rough diamonds, the percentage of the diamond which is clean is checked. The more inclusions in the stone, the lower the quality grade will be. Near gem quality stones will either be polished for the jewellery market or used for industrial purposes such as in drill bits and sawing blades.

*Boart quality diamonds at DTCB (DTCB, 2006)*

# Sorting and Valuation

*Green diamonds working sample at BDVC (BDVC, 2002)*

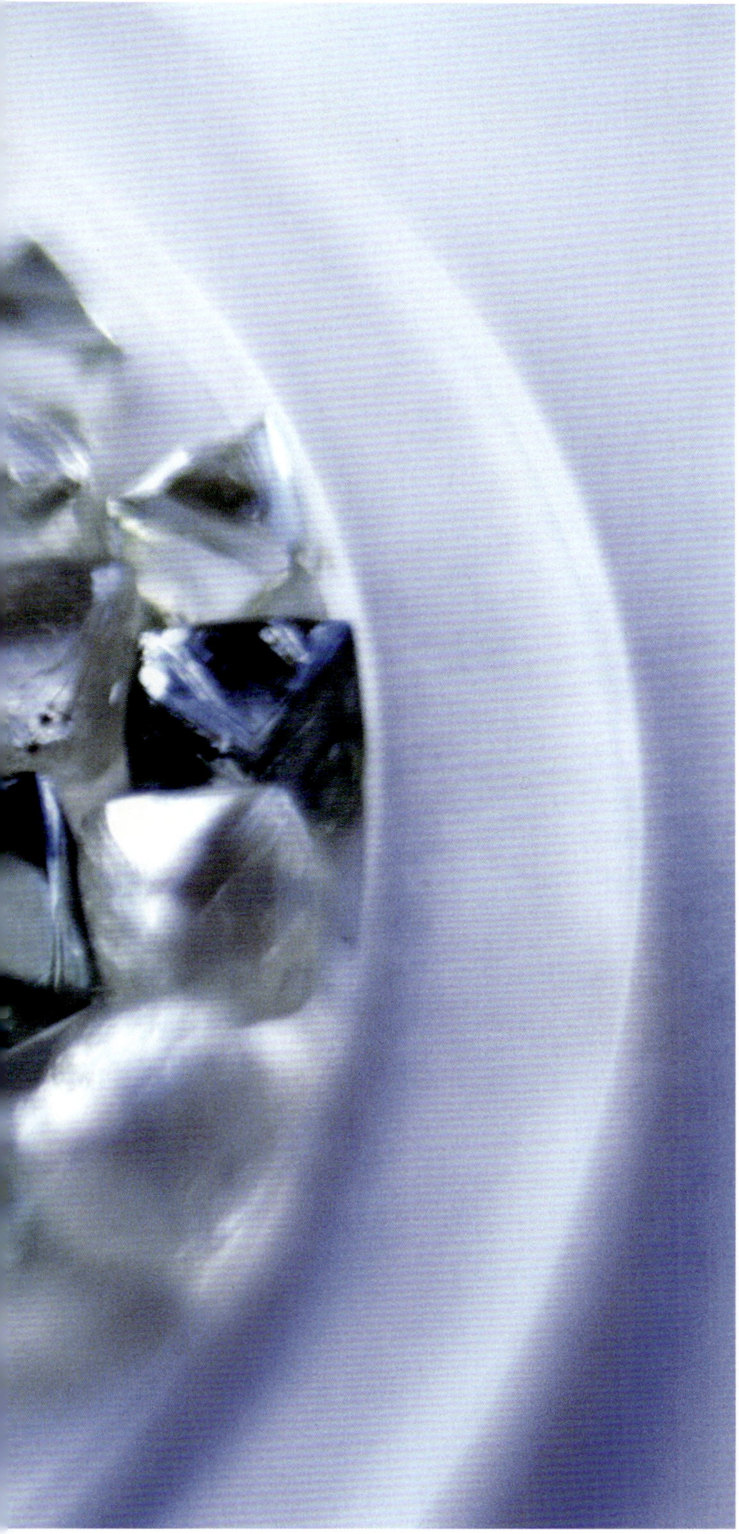

The final step requires the diamond sorter to decide on the colour category. The most common colours of diamonds fall into categories from colourless to yellow and brown, although some rarer colours such as blue, orange, pink and red do appear from time to time. These very rare examples are referred to in the industry as Fancy Coloured diamonds, but these stones are rare in Botswana. Diamonds described as 'white' are actually colourless and fully transparent. Fancy coloured diamonds are generally worth much more than white diamonds due to their rarity and beauty.

At DTCB in Gaborone, quality assurance of the sorting and valuation process is carried out by a Global Quality Assurance Department. Accurate pricing and final presentation of the diamond shipment sold by DTCB is reliant upon the quality and repeatability of the rough diamond sorting.

*Multi-coloured gems at BDVC (BDVC, 2002)*

# Sorting and Valuation

*Single medium sized rough Octahedron (BDVC)*

It is important to supply Sightholders with a consistent assortment of rough diamonds for their factories on a regular basis to help both their production process and downstream polished diamond commitments.

The actual valuation given to a single diamond is dependent on its quality as described above and each classification of diamond is given a "price" in a diamond price book which is updated on a regular basis.

Once diamonds from the individual producer countries have been valued they are carefully blended with like for like diamonds from the DTC's other producer countries to create a selling mixture. This process is called Aggregation. The DTC currently has around 110 selling mixtures or boxes.

*The three main diamond size categories, DTC (De Beers, 2007)*

# Sorting and Valuation

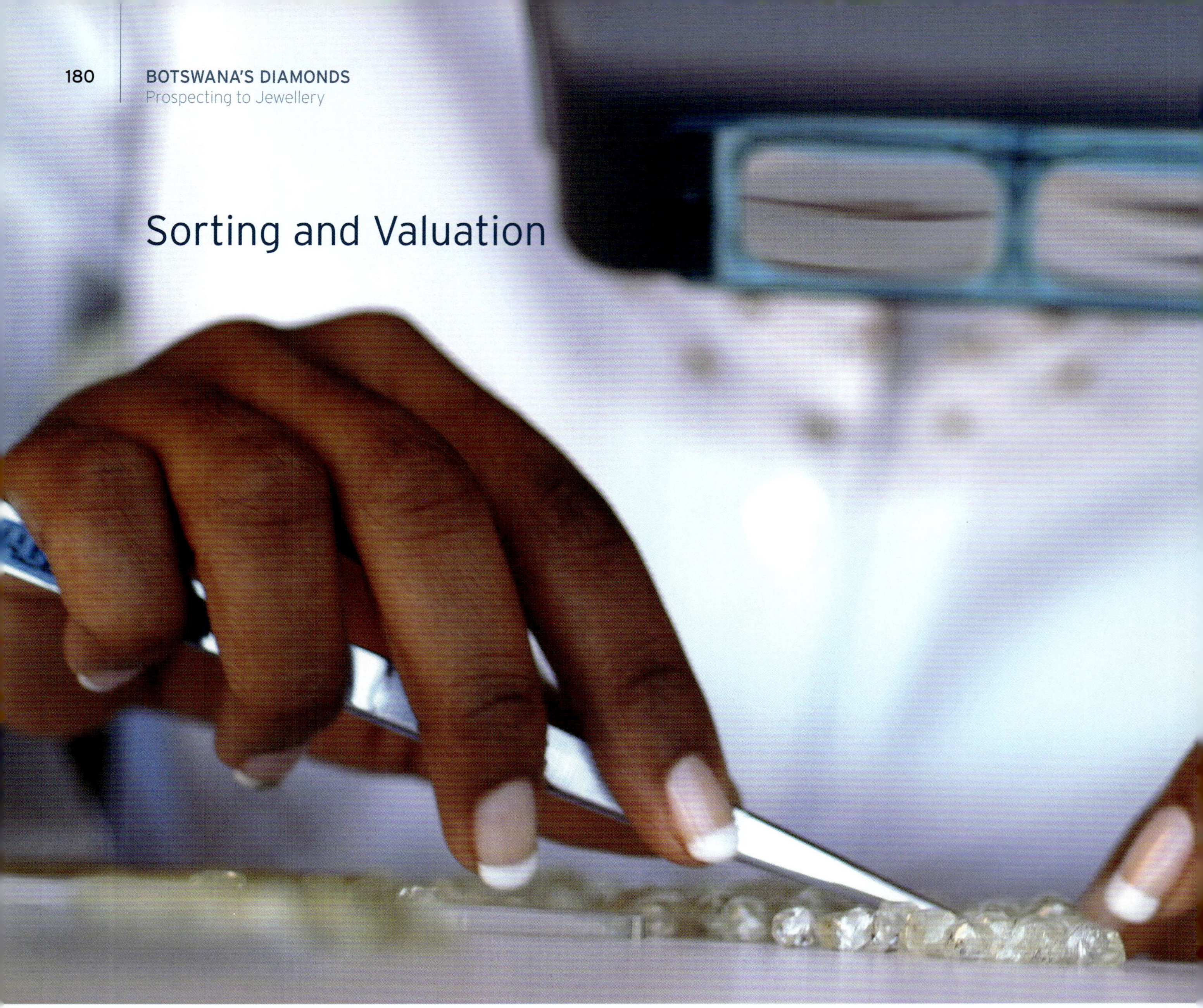

*Sorting small gems at Orapa House, Gaborone (BDVC, 2007)*

Although much of DTC's sorting process is carried out by hand using traditional handloupes and headloupes, DTC has used technology to assist with the sorting of smaller diamonds since 1972.

The DTC Research Centre, located in Maidenhead, UK, works very closely with the company's sorting and valuing offices around the world to continuously improve sorting capability, with a particular focus on enabling improved value-added assortments and simplifying processes.

The sorting technology developed over the last decade provides a strong competitive advantage for DTC and its producing partners in terms of its ability to rapidly and accurately sort large volumes of diamonds into bespoke categories of size, shape, colour, and clarity. Some of the machines can operate at speeds of up to eight stones per second.

This new technology can be found at all the partner sorting operations; in particular the technology has helped DTC Botswana become the largest and most sophisticated rough diamond sorting and valuing company in the world.

Having sorted and valued the precious gems, the next stage in the diamond pipeline is to market and sell the diamonds for downstream beneficiation.

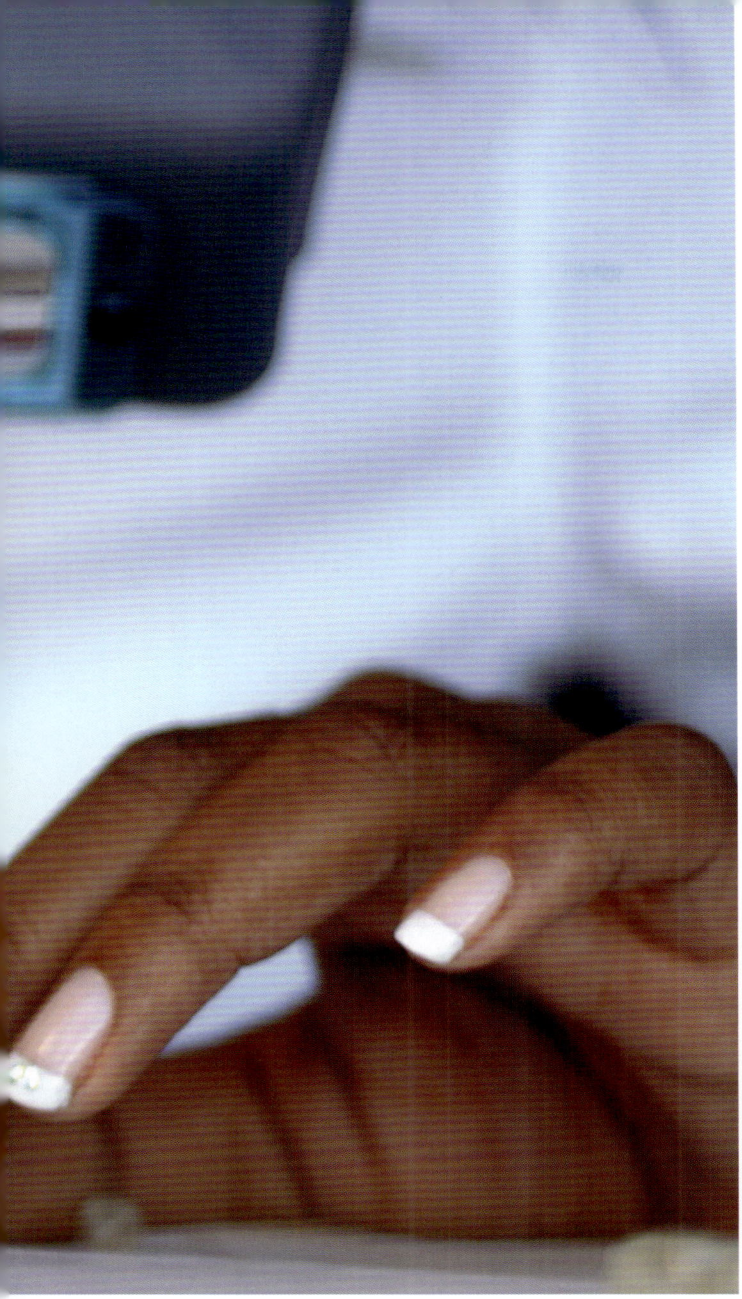

A DTC Gem Selection (De Beers, 2007)

*Aggregated small white gems at DTCB (DTCB, 2008)*

8

# Sales and Marketing

The diamond sales industry, originally established by the De Beers Family of Companies, has adopted such slogans as "Diamonds are Forever" (coined by Francis Gerety, New York, 1947) and "Diamonds are a girl's best friend" to sell diamonds since their advertisement agencies have stressed the need to strengthen the association between diamonds and romance.

In forty years, the value of diamonds has increased nearly a hundredfold. The expenditure on advertisements, whose value gradually increased to $10 million/year, has been a worthy investment.

The Diamond Trading Company (DTC) is the rough diamond distribution arm of the De Beers Family of Companies and is the world's largest supplier of rough diamonds, by value.

# Sales and Marketing

*DTCB Sorting and Valuation building (DTCB, 2008)*

With activities in sorting, valuing and diamond beneficiation, the DTC has representative offices in the UK, and South Africa, as well as joint venture operations in Botswana and Namibia with the governments of those countries. Joint ventures DTC Botswana (DTCB) and Namibia DTC – 50:50 partnerships with the governments of Botswana and Namibia – sell diamonds to clients for cutting and polishing in those countries. Formed in 1934, the DTC (formerly known as the Central Selling Organisation until the year 2000 when it was renamed DTC) was incorporated in 1986 and established as a stand-alone company within the De Beers family in July 2004. DTC's vision is to be the world's most effective rough diamond distributor maximising long-term value for Producer and Sightholder partners, while assisting in the creation of viable downstream industries in producer countries.

In Botswana today, rough diamonds are either sold through the DTCB to "sightholders", or through a much more recent development, the open tender or auction, as is the case with stones produced currently from Firestone Diamond's B/K 11 Mine in 2011 and Boteti mining's A/K 6 Karowe Mine in 2012.

# Sales and Marketing

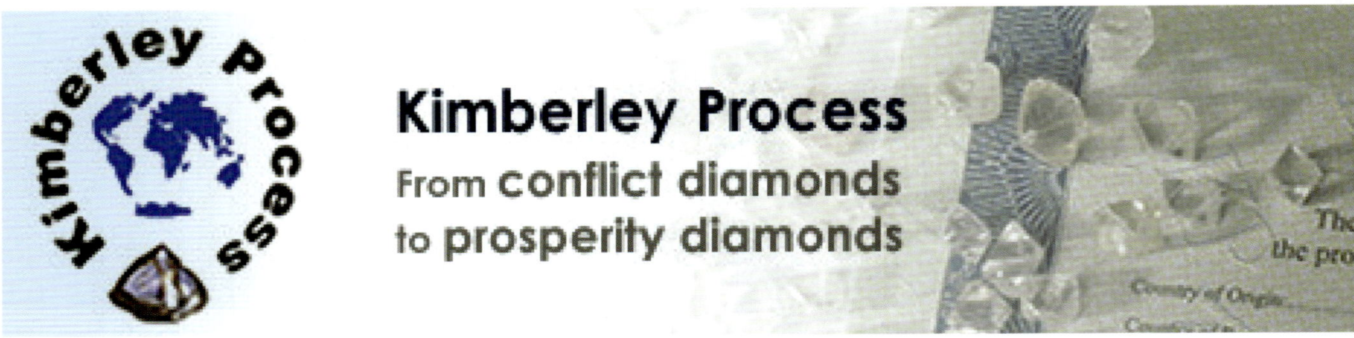


The Botswana diamond industry is fully compliant to the Kimberly Process Certification Scheme (KPCS), which was launched in May 2000 in Kimberly, South Africa. KPCS is an international initiative aimed at breaking the link between legitimate trade in diamonds and conflict diamonds. It represents a unique United Nations mandated joint initiative by approximately 75 countries, the international diamond industry and civil society to stop the flow of conflict diamonds.

Conflict diamonds are rough diamonds which are used by rebel movements or their allies to finance conflict aimed at undermining legitimate governments. The KPCS imposes extensive legally binding requirements on member states to certify that rough diamonds (both exported and imported) are conflict free. This requires that every time diamonds (rough or polished) change hands the seller must affirm on the invoice that the diamonds have been purchased through official channels and therefore not involved in conflict funding. All DTC clients subscribe to the De Beers Family of Companies Diamond Best Practice Principles (BPPs) which stipulate that all clients must adhere to the KPCS. Furthermore, 10% of all clients are audited annually to ensure that the standards are met.

*Panorama View of DTCB sort house, Gaborone (M. Brook, 2011)*

The Sales and Marketing department of DTC works as an interface between DTC Botswana and its sightholders to ensure that aggregated boxes of consistently sorted diamonds are distributed in the most efficient and effective way possible to local sightholders.

Marcus ter Haar, Head of Sales and Marketing, DTC Botswana (DTCB, 2011)

On 14th November 2007, the DTCB announced its first list of Sightholders, formalising supply arrangements for 16 Botswana-based diamond manufacturing operations. An evaluation process resulted in all 16 licence holders receiving supply and becoming DTCB Sightholders.

# Sales and Marketing

Sightholders are contracted to buy from DTCB for a 3 year period at which time they re-apply to become sightholders. The evaluation criteria seeks to promote both local value add, in terms of skills development, job creation and local manufacturing, as well as global competitiveness. The process for determining supply levels is overseen by a governance panel comprising both the Government of Botswana and De Beers representatives to ensure that it is carried out fairly.

The criteria used to determine individual supply levels included global competitiveness criteria, which assesses technical, marketing and distribution efficiency; as well as local criteria, which seek to build capacity in local manufacturing. In addition, all applicants had to satisfy certain gating criteria such as compliance with the Diamond Trading Company's Best Practice Principles. The objective assessments led to tailored supply for those successful sightholders in the application process for each category of diamonds available. The 16 successful applicants represent a solid foundation and the promise of an exciting future for Botswana's diamond manufacturing industry.

*All sight holders have to adhere to the De Beers Best Practice Principles*

# Sales and Marketing

Below is a list of DTCB Sightholders who were in full operation up until March 2012 when their current licences expired and were then renewed:

- Dalumi Botswana (Proprietary) Limited
- Zebra Diamonds (Proprietary) Limited
- Yerushalmi Bros Botswana (Proprietary) Limited
- Teemane Manufacturing Company (Proprietary) Limited
- Suashish Diamonds Botswana (Proprietary) Limited
- Steinmetz Diamonds Botswana (Proprietary) Limited
- Safdico Botswana (Proprietary) Limited
- Laurelton Diamonds Botswana (Proprietary) Limited
- Pluczenik Botswana (Proprietary) Limited
- MotiGanz Botswana (Proprietary) Limited
- Lazare Kaplan Botswana (Proprietary) Limited
- Leo Schachter Botswana (Proprietary) Limited
- Hearts & Arrows Cutting Works Botswana (Proprietary) Limited
- Eurostar Botswana (Proprietary) Limited
- Dia Holdings (Proprietary) Limited
- Shrenuj Botswana

In December 2011 DTCB announced that all existing sightholders were successful in their application renewal and that subject to Government regulatory provisions an additional five sightholders would be added to the DTCB customer list, making a total of 21.

Assuming Government awards the relevant licences the five new sightholders are:

- Blue Star Diamonds Botswana
- Julius Klein Diamonds Botswana
- Chow Tai Fook Cutting Works
- Tache Botswana
- Arabov Life Diamonds Botswana

The DTC requires that all applicants to become sightholders meet certain mandatory requirements before their application can be considered:

DTCB requires all applicants to be able to demonstrate a minimum level of turnover and total rough diamond utilisation in each of the three prior years before application date.

For DTCB this equates to $30m annual turnover and $20m annual rough diamond utilisation. The DTC requires all applicants to be able to provide consolidated or combined audited financial statements for three consecutive years prior to their application. Another requirement is that the applicant and its key individual have, since the inception of the Kimberley Process Certification Scheme (KPCS) in January 2003, not purchased, sold, exported, imported or otherwise traded in rough diamonds without the full authority of, and in compliance with, the KPCS.

The Best Practice Principles apply to the entire De Beers Family of Companies, all Diamond Trading Company Sightholders and certain third parties including major contractors of the De Beers Family of Companies.

All are required to use their best endeavours to ensure that the companies with which they work also comply with these standards.

# Sales and Marketing

The Best Practice Principles set out various required standards of conduct with regards to three main areas:

- Business Responsibilities
- Social Responsibilities
- Environmental Responsibilities

DTC Botswana announced in December 2011 that it was *"proud of the progress that the industry has made to date in Botswana and is fully committed to fulfilling the promise of the 'Diamond Decade'. In a renewed commitment to the local industry and beneficiation, sales for local manufacturing will be increasing through the new contract period. This is in line with the agreements between the De Beers Group of Companies and the Government of Botswana to further catalyse growth in the local industry with continued creation of jobs, increased local processing of rough diamonds, and the transfer of both technology and skills to Botswana"*.

Prior to confirming the offer of a DTCB supply contract, all newly qualified applicants will have their applications verified by a third party and, at DTCB's discretion, many returning Sightholders applications will also be verified. Brian McDonald, then Managing Director of DTC Botswana, said: *"DTC Botswana is incredibly committed to partnering with its Sightholders to drive demand for diamonds and support the growth of a sustainable diamond manufacturing industry in Botswana. All the qualifying applicants represent leading diamantaires who have demonstrated a passion for adding real value to the rough we sell thus further enhancing the position of diamonds as the world's leading luxury product. We are confident that, together with our Sightholder partners, we will build a 'Diamond Decade' based upon market leading skills, investment and innovation"*. In January 2012, Tabake Kobedi was appointed as the first Motswana Managing Director of DTCB.

*Brian McDonald, MD DTCB, 2007-2011 (DTCB)*

*Tabake Kobedi, appointed as first Motswana MD of DTCB, effective 19th January, 2012*

# Sales and Marketing

*Rough and cut gems (Steinmetz, 2010)*

Diamonds from DTCB are sold at ten "sights" per year. Fitch and Pooley, an independent diamond valuing company in Botswana, has held the Botswana Government Diamond Valuer (GDV) contract for many years and its responsibility is to ensure adherence to global market prices. However, in 2009, the government awarded a three year U$9.4 million (P65 million) contract to Diaval Limited for its diamond valuation services.

The main function of the GDV is to check, on behalf of the Government, the sorting and valuation of the diamond production of Debswana Diamond Company. The GDV in Botswana gives comfort to the host government that the mine production is accurately valued - that the carats produced are exported at the appropriate values set by the Diamond Price Book. *"It is clear that the job of a GDV is not confined solely to verifying compliance of the assortment to the DTC sample. It involves a range of activities that are crucial to the development of Botswana's diamond industry and the empowerment of its citizens,"* the Botswana Government has stated in the Botswana press.

As per the September, 2011 Botswana Government - De Beers 10-year agreement, all De Beers sales to sightholders will take place in Botswana instead of London, as from the end of 2013. The 2010 De Beers sales were worth $5.08bn.

Compared to DTCB's 21 Sightholders for the period 2012-2015, DTC London has 72 Sightholders. Sightholder purchases in Botswana, are therefore expected to potentially treble by the end of 2013.

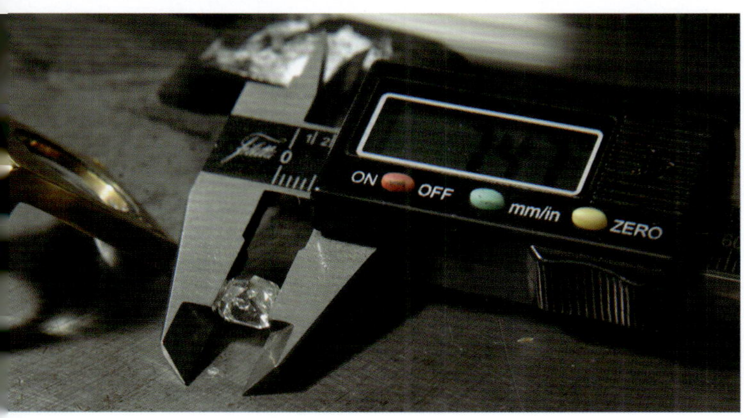


*Measuring dimensions of both rough and polished diamonds (Steinmetz, 2010)*

# Sales and Marketing

*Sorted diamonds at BDVC (BDVC, 2002)*

With the move expected to be concluded by October 2012, the first phase of the relocation of the operations of DTC International to Gaborone has gathered steam. While the agreement between government and De Beers stipulates that the relocation of DTC International from London will be completed in August 2013, DTC Botswana's then managing Director Brian McDonald says aggregation will come much sooner. *"From October next year, we should be doing aggregation of diamonds here. The whole process should see the relocation of around 80 to 90 families from London"*, he said in 2011. This means by October 2012, diamonds from De Beers mines in Namibia, South Africa and Canada will come to Gaborone for aggregation before being exported to London for the sights.

*The Miss Universe Botswana Las Vegas Diamond (Steinmetz, 2010)*

The Miss Universe Botswana Las Vegas rough Orapa diamond (Steinmetz, 2010)

*"We were prepared to move here in exchange for a 10-year agreement,"* said Bruce Cleaver, executive director for strategy and business development at De Beers, who was the chief negotiator for the company. *"We've got an arrangement where we've got what we want and the Botswana government has what it wants,"* he said.

The agreement will definitely deliver tangible outcomes and enable Botswana to achieve its aspiration to be a major diamond centre engaged in all aspects of the diamond business. In fact, the agreement also ensures the continued supply of rough diamonds to the sightholders.

The next stage in the diamond pipeline is the cutting and polishing of the rough diamonds.

# 9 Cutting and Polishing

The diamond cutting process includes the following steps; planning, cleaving or sawing, bruting, polishing, and final inspection.

Diamond manufacturers analyze diamond rough from an economic perspective, with two objectives steering decisions made about how a faceted diamond will be cut. The first objective is that of maximum return on investment for the piece of a rough diamond. The second is how quickly the finished diamond can be sold. Scanning devices are used to get a 3-dimensional computer model of the rough stone. Also, inclusions are photographed and placed on the 3D model, which is then used to find an optimal way to cut the stone.

*Diamond Technology Park Main Entrance, 2011 (M. Brook)*

Long before the Diamond Technology Park (DTP) was set up in Gaborone in 2008 Diamond Manufacturing Botswana (DMB) established Botswana's first diamond polishing plant in Gaborone in 1976 and in 1991 Teemane Manufacturing Company (Pty) Ltd. was also established. Based in Serowe, and then owned by Debswana, it employed over 200 staff and exported its production to De Beers' Diamond Trading Company (DTC). In 2003, Teemane was taken over by Diarough who continue to operate the diamond cutting and polishing factory today.

*Former Botswana President Fesus Mogae and Steinmetz CEO Ori Temkin at the inauguration of the Steinmetz diamond factory at the DTP on 6th February, 2008*

In 1992, Lazare Kaplan Botswana opened a diamond cutting and polishing factory in Molepolole, 40km west of the capital city, Gaborone, concentrating their production on smaller stones, of less than 0.18 carats. This factory, which first employed 500 Batswana ladies was taken over by Leo Schachter Botswana in 1997. Belgium's Eurostar Group constructed a new diamond cutting and polishing factory in Gaborone in 2004 and employs about 350 staff.

However, from the mid 2000s, given Botswana's keen desire to develop more aspects of diamond beneficiation within the country, it became obvious that there was quite an urgent requirement to develop a dedicated Diamond Technology Park to house all the new players that would be expected to flock to Gaborone to be part of the Diamond Dream.

*The Shrenuj Company was founded in Bombay, India in 1906 by Kalidas Doshi. Kirtilal Doshi, Chairman Emeritus (centre), Shreyas Doshi, Corporate Managing Director (left) and 4th Generation, Vishal Doshi, Group Executive Director (right). (Shrenuj, 2006)*

*Current tenants of the atrium at the Diamond Technology Park (M. Brook, 2012)*

**The table below provides details of the activities of the 16 DTCB "Sightholders" in Botswana:**

| Company | Year | Development |
|---|---|---|
| Diamond Manufacturing Botswana (Star Group) Dia Holdings | 1976 | Botswana's first diamond cutting & polishing factory, Located in Gaborone, it is now part of the London based Star Diamond Group, established in 1948. |
| Temaane Manufacturing Company (Diarough) | 1991 | The original Serowe based factory was established in 1991 and was acquired by Diarough in 2003 and is one of the largest diamond factories in Southern Africa. |
| Lazare Kaplan | 1992 | Lazare Kaplan, founded in 1903, opened a diamond cutting and polishing factory in Molepolole, 40km west of Gaborone, in 1992 concentrating on small size diamonds. |
| Leo Schachter | 1998 | Bought a diamond cutting and polishing factory in Molepolole, 40km west of Gaborone, from Lazare Kaplan in 1998, concentrating on small size diamonds. |

# Cutting and Polishing

| Company | Year | Development |
|---|---|---|
| Eurostar | 2004 | Belgium-based Eurostar occupies a 13,000m$^2$ plot in Broadhurst, Gaborone with 340 staff. |
| Suashish | 2004 | Was awarded a 10 year diamond manufacturing licence in 2005 and established a Gaborone based factory in 2007. It is the first Indian Company to be approved as a DTCB "Sightholder". |
| Southern African Diamond Corporation (SAFDICO) | 2006 | Established in Botswana in 2006, became operational in September, 2007, 120 staff, cuts and polishes diamonds from 2 to 100 cts. Designed and developed the Diamond Technology Park (DTP) which opened in 2008. |
| Pluczenik | 2006 | A Belgian-based company first established a diamond sorting facility in Gaborone in 2006 and opened a cutting and polishing factory in February 2007, employing 300 staff. In 2011, construction of a new factory commenced in the DTP. |
| Zebra Diamonds | 2007 | Part of the Belgian IGC Group established in 1912, opened a new diamond factory close to DTP on a 10,000m$^2$ plot with a capacity to employ 120 diamond polishers. |
| Dalumi | 2007 | Opened a 100m$^2$ state of the art diamond polishing factory in Mogoditshane, close to Gaborone, with more than 200 staff, the factory input is 5,000 cts per month. |

| Company | Year | Development |
|---|---|---|
| Motiganz<br>MOTIGANZ DIAMOND GROUP | 2007 | First established in Botswana in 2006, and specialising in "cubes", a 1500m² factory was opened in Gaborone in April 2007, employing 300 staff. |
| Hearts and Arrows<br>EXELCO GROUP | 2007 | Part of the Exelco Group, opened a factory close to DTP in 2008 and is the sister factory to the H & A Bangkok factory in Thailand. |
| Yerushalmi<br>YERUSHALMI BROS DIAMONDS | 2007 | Israel-based, established a large scale highly automated diamond factory in Mogoditshane in 2007. |
| Steinmetz<br>Steinmetz<br>Creators of the world's finest diamonds | 2008 | Part of the 70 year old Steinmetz Group, opened a state of the art diamond cutting and polishing factory in the DTP in February 2008. |
| Shrenuj<br>SHRENUJ<br>HERITAGE. INNOVATION. PASSION | 2010 | Part of the 105-year-old Shrenuj Group which originated in Bombay, India, has a factory in Broadhurst, Gaborone and a diamond Jewellery unit in the Diamond Technology Park. |
| Laurelton<br>LAURELTON DIAMONDS BOTSWANA<br>A SUBSIDIARY OF TIFFANY & CO. | 2011 | First established in Botswana in 2007, a wholly-owned subsidiary of Tiffany & Co, opened a US$ 5 million diamond cutting and polishing factory in Gaborone in December, 2011, occupying a plot of 3,000m² and employing 140 people. |

# Cutting and Polishing

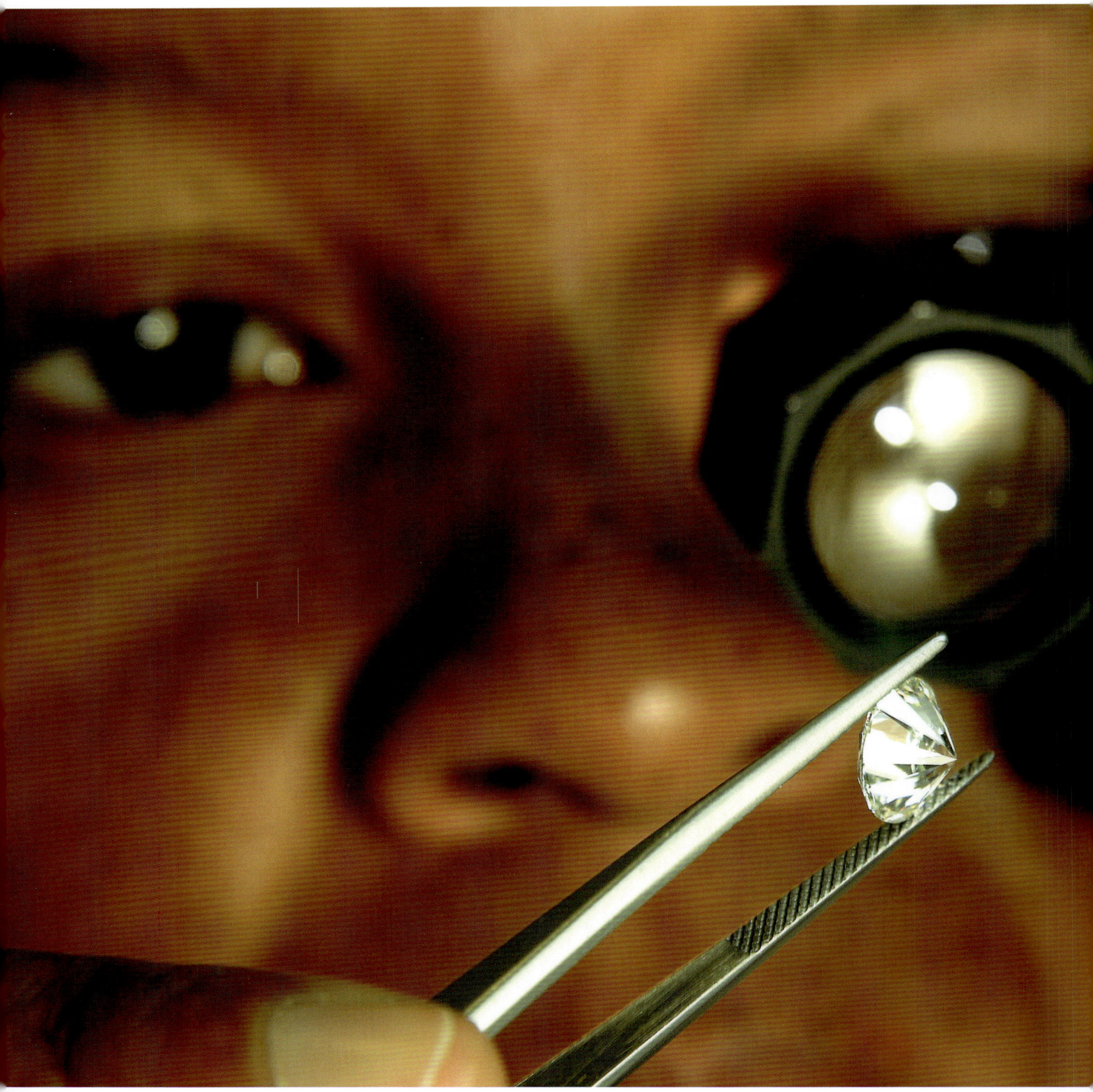

*A beautiful cut and polished round gem (Steinmetz, 2010)*

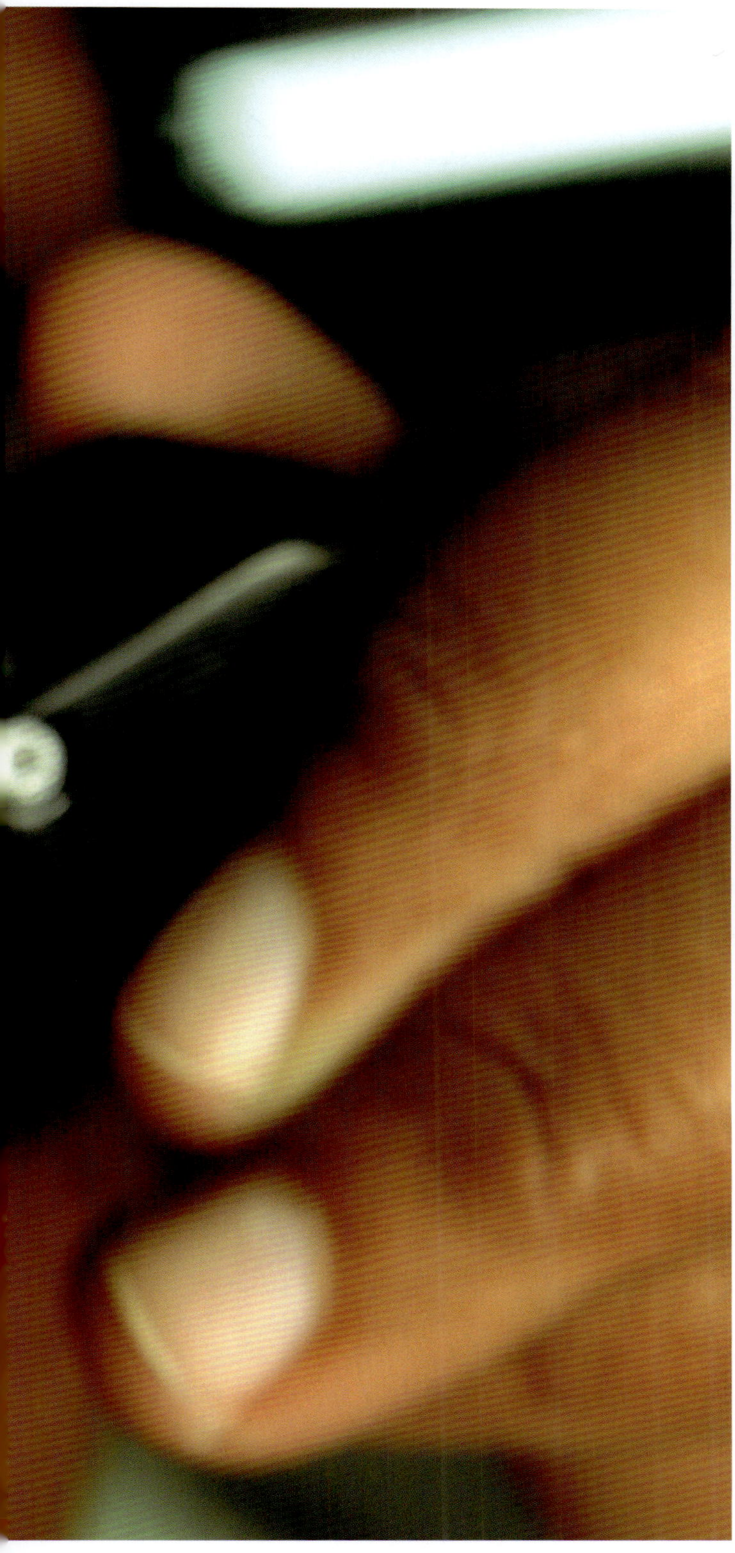

Diamond cutting in Botswana is regulated by the Diamond Cutting Act, Chapter 66:04, of 1979 which controls the cutting, sawing, cleaving and polishing of rough and uncut diamonds. Companies operating in this business in Botswana must have a valid Diamond Cutter's licence under the terms of the Diamond Cutting Act and a Precious Stone Dealer's licence under the terms of the Precious and Semi-Precious Stones Act. The Act requires every manufacturer to *"keep a true and correct daily record, in the prescribed form, of such particulars of all his dealings in diamonds and all his diamond cutting operations..... Every licencee shall within 14 days after expiration of every month forward to the Mining Commissioner a true copy of all such entries in such record for the previous month, together with a solemn declaration testifying the correctness thereof".*

The regulatory body for monitoring compliance with the above-mentioned Act is the Diamond Office, which is part of the Mineral Affairs division of the Ministry of Minerals, Energy and Water Resources. Currently housed in the Diamond Technology Park next to the Government Diamond Hub, the main activities of the Office are:

1) The inspection of both rough and polished stones at the diamond dealers offices
2) Supervision of the import and export of rough and polished diamonds
3) Issuance of Kimberly Process Certificates to diamond cutting factories
4) Issuance of cutting and diamond dealing licences

# Cutting and Polishing

*Employee training on principles of diamond cutting (Steinmetz, 2010)*

The official launching of the DTP was performed by the Hon. Minister of Minerals, Energy and Water Resources, Dr. P.H.K. Kedikilwe on 28th January, 2009. The Minister said "*This facility is in line with our vision of taking Botswana's diamond industry to a different level, a vision that we have been pursuing since 2004 and a vision that we intend to pursue even in these difficult times.*

*Our vision is, together with development partners like Safdico and many others represented here, to transform Botswana from being not only a diamond mining giant, but to also become a world class diamond centre. We know and we have always known that this will not be an easy task; our commitment to delivering on this vision is however unwavering*".

*The Botswana Diamond Manufacturing Factory in Gaborone (Diamond Manufacturing Botswana, 2011)*

*A factory worker inspecting a Polished stone (Diamond Manufacturing Botswana, 2011)*

He further went on to say "*I need not remind anyone here about the importance of diamonds to Botswana's economy. Diamonds have contributed and continue to make significant contribution to Botswana's economy. Export earnings and government revenues are still dominated by diamonds. It is therefore no surprise that we become seriously concerned when diamond markets go through what they are going through. As the world's largest diamond producer by value, it is both fitting and relevant that Botswana develops a thriving diamond centre, creating among others, employment, skills transfer, beneficiation, diamond support services and optimal use of our natural resources*".

The DTP is a private initiative funded and headed by the Safdico Group, a Diamond Trading Company Sightholder, with offices and polishing facilities in Johannesburg, Antwerp, London, Mauritius, Geneva, New York and Botswana. Programs of diamond cutting and polishing have now been transferred to Botswana for the advancement of local beneficiation and skills development in the DTP.

# Cutting and Polishing

*Inspecting a cut and polished stone (Steinmetz, 2010)*

Safdico Botswana's diamond cutting facility became operational in September 2007. The facility is state of the art and utilizes world-class equipment and technology. Safdico employs approximately 100 people. Batswana are trained to polish diamonds of sizes ranging from +2carats to +100carats according to stringent international standards. DTP houses diamond manufacturers as well as those servicing the diamond industry. These businesses are positioned in a secure and efficient environment, in the same neighbourhood as DTCB and the Debswana Diamond Company Corporate Centre.

DTP's buildings are architecturally relevant and beautiful. Tenants of the central services building (known as the Atrium) include: the Ministry of Minerals, Energy and Water Resources (the Diamond Hub), the Gemological Institute of America, ABN-Amro Bank, First National Bank, Stanbic Bank, Malca Amit, Brinks, Sarin, Ogi, Bonas, Hennig and Mining Companies Boteti and Firestone Diamonds.

Tenants of the mini factory building include Shrenuj and Private Collections and a central scouring facility.

The Botswana Police operate a satellite police station on site which is armed and operational 24/7, to support tenants and insurers. DTP has also built a zebra enclosure intended to maintain the theme of making Botswana's diamond centre unique. The zebra, Botswana's national animal flies the flag for the world's number one producer of diamonds. This significance is characterised in the DTP logo which was inspired by the zebra.

# Cutting and Polishing

*Setting up for diamond polishing (Shrenuj, 2011)*

In addition to putting up its own world-class diamond cutting facility, Safdico has played a pivotal role in the establishment of the Botswana diamond industry. It has provided an enabling and secure operating environment. It has invested significantly in the Botswana economy, encouraged foreign direct investment and positioned Botswana in its rightful place on the international map with a state of the art diamond centre.

The Gemmological Institute of America (GIA) was established in the Botswana DTP in 2008. With a tradition of science and education, the GIA was founded in 1931 and is the world's foremost authority on diamonds, colored stones, and pearls. The 81-year old organization created the International Diamond Grading System™ and the 4Cs (Color, Cut, Clarity and Carat weight), and graded the Hope Diamond, Taylor-Burton Diamond and Centenary Diamond. The issue of a diamond grade certificate is not compulsory, but many buyers now insist on the report when they purchase.

As an independent organization, GIA's mission is to protect the public trust in gemstones through its education programs, laboratory services and research findings. In addition to a multitude of online courses through GIA's eLearning programs, the education arm has 11 campuses in nine cities around the world, as well as satellite courses in countries that include Botswana. GIA has educated more than 300,000 students worldwide.

# Cutting and Polishing

GIA is the world's most respected gemmological laboratory. Since the 1930s, GIA researchers have made numerous breakthrough contributions to the public's understanding of gems. These milestones include:

- creating the D-to-Z colour scale and Flawless-to-$I_3$ clarity scale for diamonds (1953), internationally recognized standards for evaluating diamond quality
- detecting irradiated yellow diamonds (1956)
- the first report on faceted synthetic diamonds (1971)
- detecting fracture-filled diamonds (1989, 1994)
- distinguishing natural from synthetic diamonds (1995)
- identifying the effect of fluorescence on diamond appearance (1997)
- creating a comprehensive system for grading diamond cut quality (2004)

GIA conducts classes in Botswana, and has operated a laboratory for three years.

As of February 2011, GIA Education programs offered in Botswana are accredited by the Botswana Training Authority, a regulatory body of the Government of Botswana. GIA's Diamond Grading Lab classes are provided in Gaborone, Botswana and are designed to enhance the country's goals of increasing local employment and added value services from its diamond production.

In addition, GIA is investing in diamond grading in Botswana and a substantial amount of that investment is going towards employment and training of citizens. GIA has had a presence in the region for the last decade, conducting distance/ extension courses on diamond and gemstones in Madagascar, South Africa and Botswana.

*Shrenuj Jewellery Factory at the Diamond Technology Park (Shrenuj, 2011)*

## CARAT

The carat, the standard unit of weight for diamonds and other gemstones, takes its name from the carob seed. Because these small seeds had a fairly uniform weight, early gem traders used them as counterweights in their balance scales. The modern metric carat, equal to 0.2 grams, was adopted by the United States in 1913 and other countries soon after.

*A rough stone being processed in the planning stage on a planner computer (Steinmetz, 2010)*

## COLOR

After size, colour is the first step to the pricing of a diamond, and as with the rough, the whiter colours are the most sought after. Fancy colours are in a different set of price bands altogether and are affected by the saturation of the colour as well as colour overtones such as orangey-pink.

Before GIA developed the D-Z Color Grading Scale, a variety of other systems were loosely applied. The result of all these grading systems was inconsistency and inaccuracy. Because the creators of the GIA Color Scale wanted to start from scratch, without any association with earlier systems, they chose to begin with the letter D—a letter grade normally not associated with top quality.

Today, a carat weighs exactly the same in every corner of the world. The size of the polished gem does matter and is further complicated by reaching stipulated price bands, such as 0.25ct, 0.50ct, 1.00ct, 3.00ct, 5.00ct, and 10.00ct and so on. The price difference between a 0.99 carat and a 1.00 carat is very high and this continues throughout all the size bands.

# Cutting and Polishing

*Close-up view of an internally flawless round brilliant cut diamond (GIA, 2011)*

*A Crystal inclusion in a Diamond (GIA, 2011)*

*GIA's Junior Gemologist program in Botswana. Students from Tshwaragano Primary School, Ikageng Primary School, Legae Academy and Mogoditshane Community Junior Secondary School participated in the program in Gaborone. (GIA, 2012)*

## CLARITY

The internal quality is very important and there are large price differentials between internally flawless diamonds and those with flaws. Therefore, the higher the clarity, the more prized and expensive the polished gem becomes.

Diamond clarity refers to the absence of any inclusions or blemishes. Diamonds without these birthmarks are rare, and rarity affects a diamond's value. Using the GIA International Diamond Grading System™, diamonds are assigned a clarity grade that ranges from flawless (FL) to diamonds with obvious inclusions (I3). Like the color scale, GIA's clarity grading system developed because jewelers were using terms that were easily misinterpreted. Today, even if you buy a diamond in another part of the world, the jeweller will likely use terms such as VVS1 or SI2.

GIA
GEMOLOGICAL INSTITUTE OF AMERICA®

5355 Armada Drive | Carlsbad, CA 92008-4602
T: 760-603-4500 | F: 760-603-1814

**GIA Laboratories**
Bangkok Carlsbad Gaborone
Johannesburg Mumbai New York

www.gia.edu

GIA REPORT 6107487215

## DIAMOND GRADING REPORT

December 17, 2008

Laser Inscription Registry .......... **GIA 6107487215**
Shape and Cutting Style ............ **Round Brilliant**
Measurements .................. 8.13 - 8.17 x 5.10 mm

GRADING RESULTS - GIA 4CS

Carat Weight .................................... **2.07 carat**
Color Grade ................................................ **F**
Clarity Grade ........................................... **VS1**
Cut Grade ......................................... **Excellent**

ADDITIONAL GRADING INFORMATION

Finish
Polish ............................................ Very Good
Symmetry ......................................... Excellent
Fluorescence .................................................. None
Comments:
*SAMPLE *SAMPLE *SAMPLE *SAMPLE
Additional Inscription:
YOUR LOGO OR PERSONAL MESSAGE HERE

REFERENCE DIAGRAMS

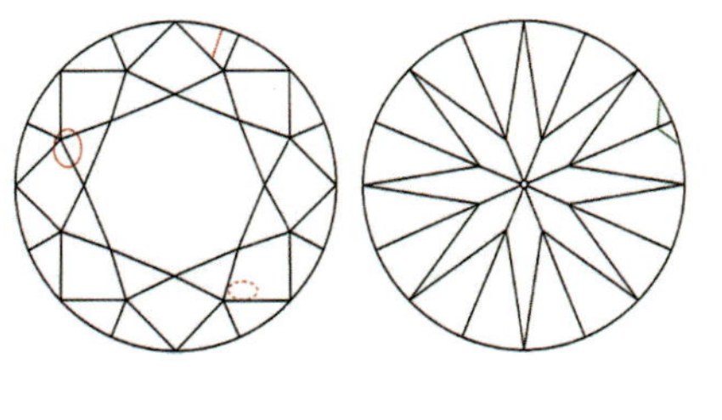

KEY TO SYMBOLS
Crystal
Cloud
Feather
Natural

Red symbols denote internal characteristics (inclusions). Green or black symbols denote external characteristics (blemishes). Diagram is an approximate representation of the diamond, and symbols shown indicate type, position, and approximate size of clarity characteristics. All clarity characteristics may not be shown. Details of finish are not shown.

GIA COLOR SCALE
COLORLESS: D, E, F
NEAR COLORLESS: G, H, I, J
FAINT: K, L, M
VERY LIGHT: N, O, P, Q, R
LIGHT: S, T, U, V, W, X, Y, Z

GIA CLARITY SCALE
FLAWLESS
INTERNALLY FLAWLESS
VERY VERY SLIGHTLY INCLUDED: $VVS_1$, $VVS_2$
VERY SLIGHTLY INCLUDED: $VS_1$, $VS_2$
SLIGHTLY INCLUDED: $SI_1$, $SI_2$
INCLUDED: $I_1$, $I_2$, $I_3$

GIA CUT SCALE
EXCELLENT
VERY GOOD
GOOD
FAIR
POOR

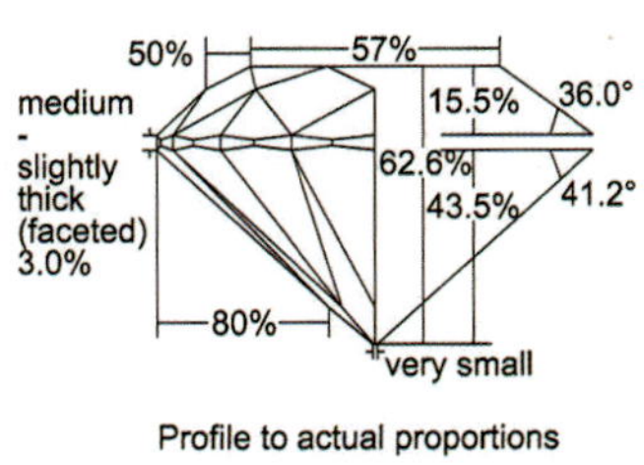

Profile to actual proportions

210100777776

This Report is not a guarantee, valuation or appraisal and contains only the characteristics of the diamond described herein after it has been graded, tested, examined and analyzed by the laboratory providing this Report ("GIA") and/or has been inscribed using the techniques and equipment used by GIA at the time of the examination and/or inscription. Inscriptions reported in this document are not a guarantee, validation, or warranty of a diamond's quality, country of origin or source; or that the diamond will be identifiable by the inscription in the future (since inscriptions can be removed). GIA makes no representation concerning any trademark, word, or symbol which is inscribed by GIA or which is identified on this Report. The recipient of this Report may wish to consult a credentialed jeweler or gemologist about the information contained herein.

The security features in this document, including the hologram, security screen and microprint lines, in addition to those not listed, exceed document security industry guidelines.

**IMPORTANT LIMITATIONS ON BACK**


*Sample GIA Diamond Grading Report (GIA, 2011)*

# Cutting and Polishing

*Inspecting a diamond after polishing (Shrenuj, 2010)*

## CUT

Rounds are the most desired cut or shape, followed by fancy cuts such as the Princess and Pear cuts. Fancy cuts are always discounted to the round shape and this decreases according to the cut with shapes such as the Heart and Trilliant as well as some obscure shapes having the highest discount.

Though extremely difficult to analyze or quantify, the cut of any diamond has three attributes: brilliance (the total light reflected from a diamond), fire (the dispersion of light into the colors of the spectrum), and scintillation (the flashes of light, or sparkle, when a diamond is moved). In 2005, GIA unveiled a diamond cut grading system for standard round brilliants in the D-to-Z color range. This system, the product of more than 15 years of intensive research and testing, assigns an overall diamond cut grade ranging from Excellent to Poor.

The quality of the cut in terms of actual polishing such as Cut Grade, Polish and Symmetry is essential in achieving the incredibly rare clarity rating of Flawless gems. Therefore, it is essential that if you have the correct size, for example a 1.05 carat polished stone, it should be a D colour round brilliant shape diamond that has no internal or external blemishes and be graded Excellent for the Cut Grade, Polish and Symmetry. The stone should have zero fluorescence.

As with rough diamonds, there are other factors which influence the overall gem description which include white or fancy colour diamonds that have high blue fluorescence or stones with unattractive clouds, milkiness or colour banding/zoning.

# Cutting and Polishing

A diamond grade report shows a laser inscription registry number, shape and cutting style, the diamond's measurements (length, breadth and width), carat weight, colour grade, clarity grade, cut grade and finished polish, symmetry and fluorescence.

The establishment of ABN AMRO Bank within DTP, which will service the diamond industry, is another major step towards making Botswana an international diamond hub. Speaking at the bank's opening ceremony at the DTP in 2010, Vice President Lt Gen. Mompati Merafhe said *"the presence of ABN AMRO Bank is most welcome as it will provide the much needed finance to fuel the downstream industry"*. ABN AMRO Bank is the 12th largest bank in the world and the number one bank servicing the diamond industry and the Vice President said by establishing its presence in Botswana it is clearly expressing its confidence in investors in the local diamond industry.

ABN·AMRO

*Bank ABN AMRO, the largest financier in the world to the diamond industry arrived in Botswana in 2009 and opened their offices in the Diamond Technology Park in 2010.*

Lt Gen. Merafhe said *"financing the diamond industry was a vital component of any emerging diamond centre and crucial for the development and success of Botswana and the Diamond Hub. Diamond activities thrive on specialised financing, and therefore for Botswana to prosper as a World Diamond Centre the presence of a bank with global financing expertise in the diamond industry is crucial"*.

The diamond and jewellery industry has been part of ABN AMRO Bank's focus for over a century now and this bank is the global market leader and oldest dedicated financial services provider to the industry.

As part of its continuous focus on creating value for clients and stakeholders, the bank has a dedicated team of specialists: the International Diamond and Jewellery Group (ID&JG) which focuses on commercial banking and has a strong position in the industry's key centers such as Antwerp, New York and Mumbai.

By extending its footprint into Botswana in 2009, ABN AMRO demonstrated its commitment to support Botswana's dream to establish a downstream diamond industry and a regional trading hub for diamonds. ABN AMRO was therefore a first mover in the diamond banking landscape of Botswana and plays an essential role in the further development of the diamond industry, as there are few banks that specialize in financing the working capital needs of diamond companies.

*Preparing to polish (Steinmetz, 2010)*

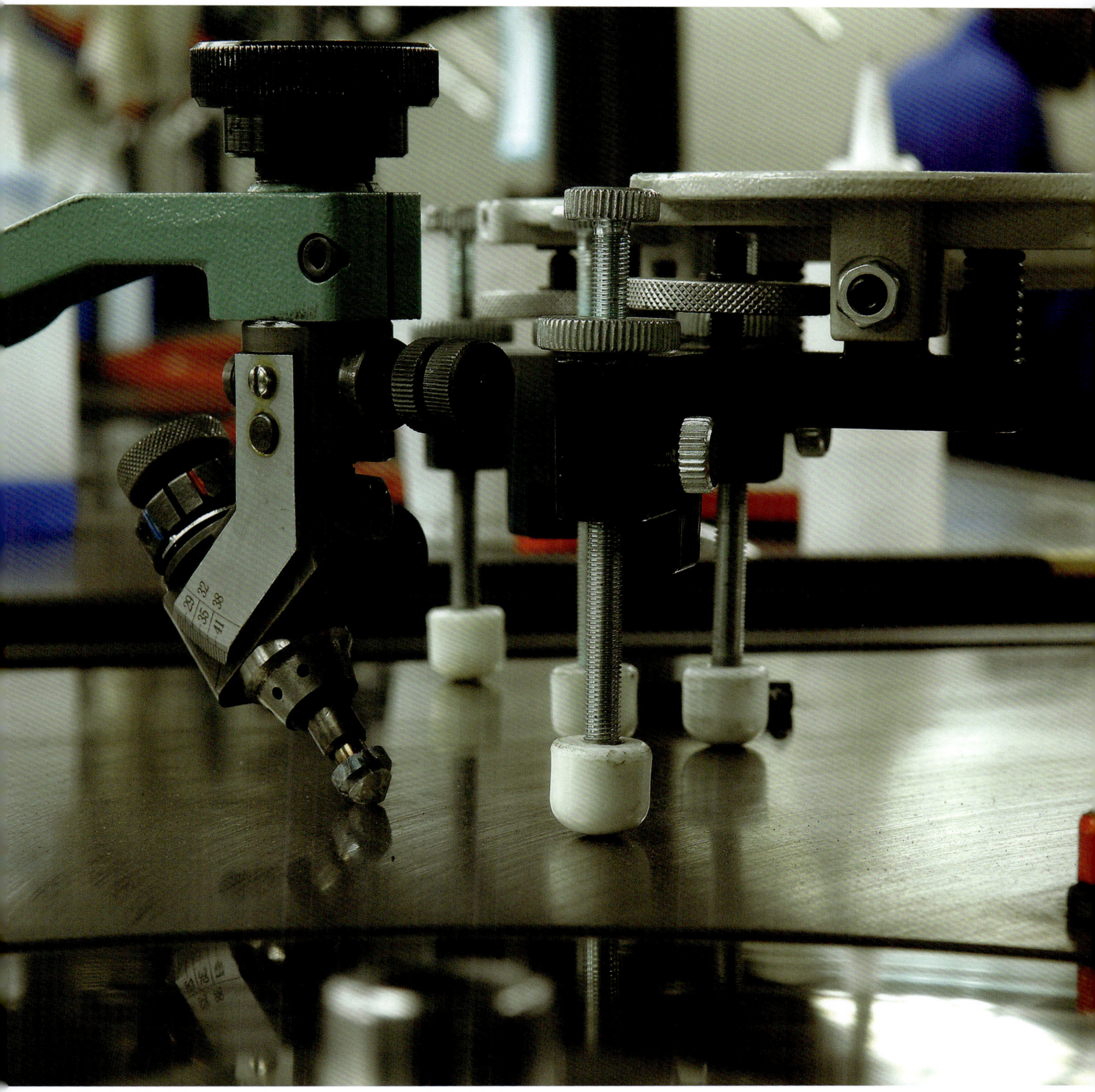

*A rough diamond in preparation stage (Steinmetz, 2010)*

# Cutting and Polishing

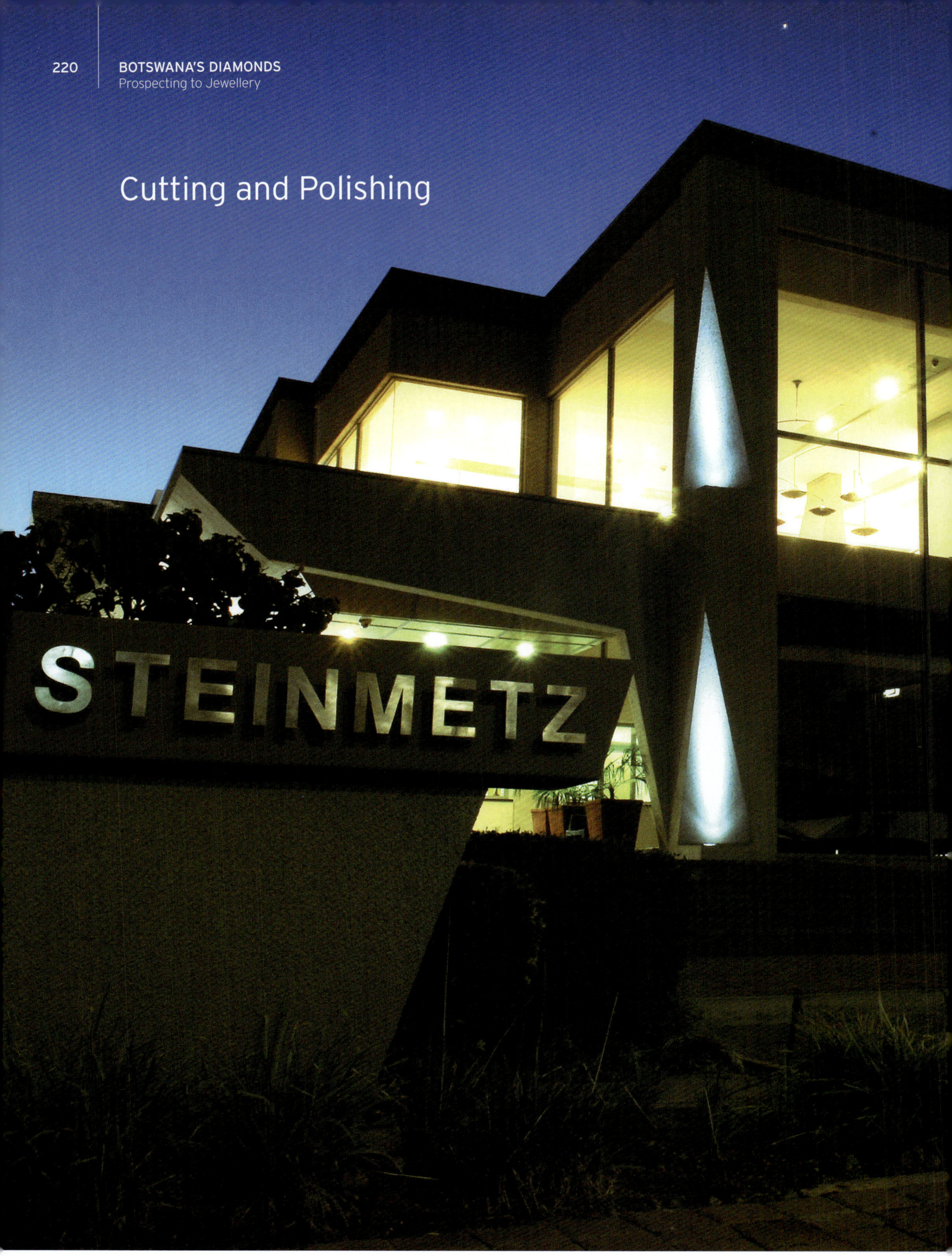


*Steinmetz Headquarters, DTP (Steinmetz, 2010)*

*Orapa Mine rough diamond ready for cutting and polishing (Steinmetz, 2010)*

*Inside the Steinmetz diamond cutting factory, Gaborone DTP (Steinmetz, 2010)*

# Cutting and Polishing

*Shrenuj cutting and polishing factory Broadhurst, Gaborone (Shrenuj, 2011)*

Diamond cutting and polishing takes place in factories both within and outside of the DTP. Diamond cutting is the art, skill and, increasingly, science of changing a diamond from a rough stone into a faceted gem. Cutting diamond requires specialised knowledge, tools, equipment, and techniques because it is an extremely delicate process. The first guild of diamond cutters and polishers (diamantaires) was formed in 1375 in Nuremberg, Germany, and led to the development of various types of 'Cut'. This has two meanings in relation to diamonds. The first is the shape: square, oval, and so on. The second relates to the specific quality of cut within the shape, and the quality and price will vary greatly based on the cut quality. Since diamonds are very hard to cut, special diamond-bladed edges are used to cut them.

The process of maximizing the value of finished diamonds, from a rough diamond into a polished gemstone, is both an art and a science. The choice of cut is influenced by many factors. Market factors include the exponential increase in the value of diamonds as weight increases, referred to as weight retention, and the popularity of certain shapes amongst consumers. Physical factors include the original shape of the rough stone, and location of the inclusions and flaws to be eliminated.

The weight retention analysis studies the diamond rough to find the best combination of finished stones as it relates to per carat value. The round brilliant cut and square brilliant cuts are preferred when the crystal is an octahedron, as often two stones may be cut from one such crystal. Oddly shaped crystals, such as macles, are more likely to be cut in a fancy cut—that is, a cut other than the round brilliant—which the particular crystal shape lends itself to.

# Cutting and Polishing

*Diamonds being prepared to be sawed by laser (Steinmetz, 2010)*

Even with modern techniques, the cutting and polishing of a diamond crystal always results in a dramatic loss of weight, about 50%. Sometimes the cutters compromise and accept lesser proportions and symmetry in order to avoid inclusions or to preserve the weight. Since the per-carat price of a diamond shifts around key milestones (such as 1.00 carat), many one-carat (200 mg) diamonds are the result of compromising Cut quality for Carat weight. Some jewellery experts advise consumers to buy a 0.99 carat (198 mg) diamond for its better price or buy a 1.10 carat (220 mg) diamond for its better cut, avoiding a 1.00 carat (200 mg) diamond, which is more likely to be a poorly cut stone.

In colored diamonds, cutting can influence the color grade of the diamond, thereby raising its value. Certain cut shapes are used to intensify the color of the diamond. The radiant cut is an example of this type of cut.

Natural green colour diamonds often have merely a surface colouration caused by natural irradiation, which does not extend through the stone. For this reason green diamonds are cut with significant portions of the original rough diamond's surface (naturals) left on the finished gem. It is these naturals that provide the color to the diamond.

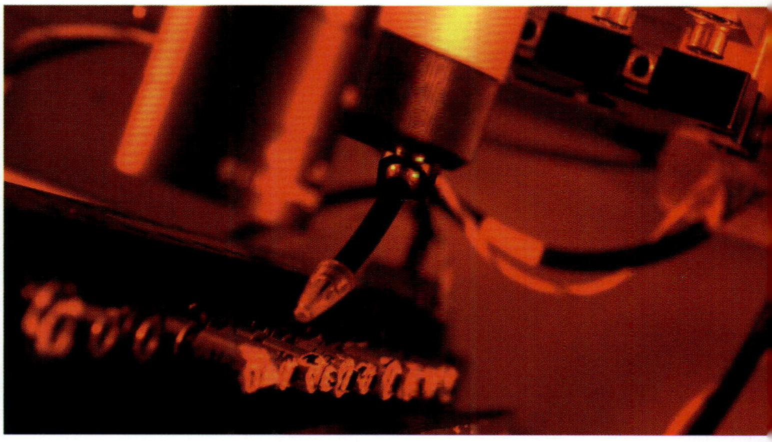

*Laser sawing a rough Diamond (Steinmetz, 2010)*

*Setting a diamond for cutting at Shrenuj Factory (Shrenuj, 2011)*

*Inspecting a bruted diamond (Diamond Manufacturing Botswana, 2011)*

*Preparing rough diamonds for bruting (Steinmetz, 2010)*

The other consideration of diamond planning is how quickly a diamond will sell. This consideration is often unique to the type of manufacturer. While a certain cutting plan may yield a better value, a different plan may yield diamonds that will sell sooner, and thereby returning the investment sooner.

The first step of cutting and polishing a diamond is to study the rough in great detail, which is a task for a rough diamond specialist or a "Marker". He takes into account the structure of the stone and the impurities that are still in it (Boonen, S & Heens, C, undated).

Cleaving is the separation of a rough diamond into separate pieces, to be finished as separate gems. Cleaving gives the rough diamond its first shape. All irregularities of the crystal and impurities are removed at this stage. The diamond is usually divided into two or more pieces. Today cleaving is usually done by laser.

Sawing is the use of a diamond saw or laser to cut the diamond rough into separate pieces. When a diamond is sawn, the stone first needs to be positioned and with a special paste a stone is affixed in a dop (holder). The diamond cutter mounts the dop on a sawing machine or uses a laser. The saw is an extra thin blade coated with a mixture of oil and diamond powder and the rotational velocity of the disc is on average 15,000 revolutions per minute. Today, lasers are being used more often for sawing diamonds.

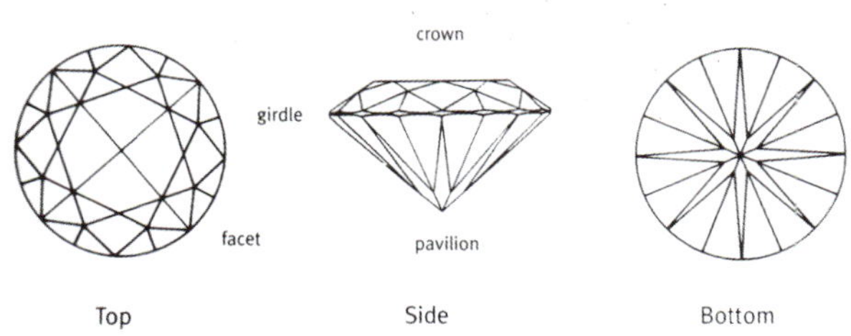


*Brilliant Diamond cross section (Boonen, S & Heens, C, undated)*

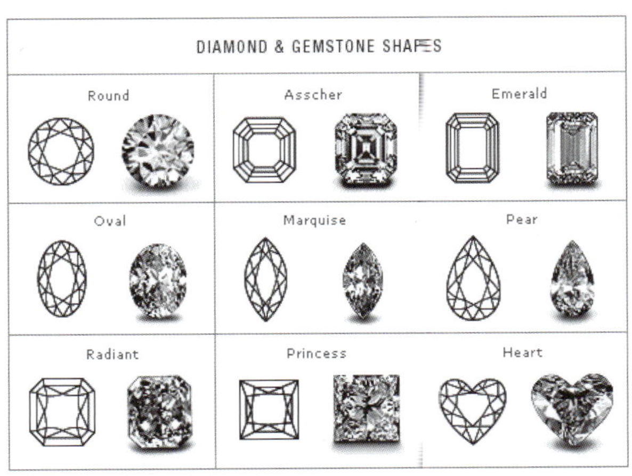


*Diamond shapes (source, Barkevs-www.barkevs.com)*

# Cutting and Polishing

*Polishing the top of a rough diamond (Steinmetz, 2010)*

*Inspecting a rough diamond prior to cutting (Steinmetz, 2010)*

Bruting is the process whereby two diamonds are set onto spinning axles turning in opposite directions, which are then set to grind against each other to shape each diamond into a round shape. This can also be known as girdling.

Polishing is the name given to the process whereby the facets are cut onto the diamond and final polishing is performed. The process takes the steps of blocking, faceting, also called "brillianteering", and polishing. Now the diamond gets its final shape and brilliance. Firstly, 4 facets are polished on the upper part (the crown) and the lower part (the pavilion) of the diamond. The result is termed "cross work". The most famous cut is the "brilliant cut" which always consists of 57 facets: besides the table, which is the large, flat facet on the top of the diamond, it has 32 facets on the crown and 24 on the pavilion.

There are nine basic shapes to cut diamonds, namely round, oval, radiant, asscher, marquise, princess, emerald, pear and heart.

The final stage involves thoroughly cleaning the diamond in acids, and examining the diamond to see whether it meets the quality standards of the manufacturer.

In 2010, Miss Universe Botswana, Tirelo Ramasedi, took a diamond to Las Vegas for the nation's representation at the Miss Universe Beauty Pageant in a couple of years. In 1999, Botswana was first represented at this prestigious pageant, held in Trinidad and Tobago, by Mpule Kwelagobe who went on to win Miss Universe and became the first black African woman to win an international beauty pageant.

Tirelo says she was also fortunate enough to visit the Diamond Empowerment Fund (DEF) which is the brain child of hip hop mogul Russell Simmons. Founded in 2007 and transforming Africa through education, the DEF is a non-profit international organisation with the mission to raise money to support education initiatives that develop and empower economically disadvantaged people in African nations where diamonds are a natural resource. The DEF has directly empowered over 1,800 students throughout its three beneficiaries based in Africa: CIDA City Campus, African Leadership Academy (ALA), and Botswana Top Achievers.

As required by the pageant all participants had to bring a national gift and Tirelo took the backbone of this country's success, a diamond. *"I called it the 'Botshelo (life) diamond"*, she said, *"Being the world's biggest producer by value our nation's economy is based on the diamond and as such it has given us life,"* Tirelo related. The diamond apparently stole many hearts becoming the most high profile gift with attendants at the auction wanting to touch and feel the diamond.

The 1.66 carat Orapa-mined diamond, cut and polished by Steinmetz Botswana in just 12-hours, just in time for Tirelo to catch her flight, was auctioned in Las Vegas at the gala dinner for $16,000, which was donated to the Nevada Cancer Institute.

In late 2011, yet another diamond manufacturing plant was opened in Gaborone, Botswana. Laurelton Diamonds, a subsidiary of Tiffany & Co. and a DTCB sight holder opened a newly-constructed 3,000 square metre diamond cutting and polishing factory built at a total cost of P36.5 million while an additional P33.3 million was used for training its own workforce to cut and polish diamonds.

According to the CEO of Tiffany & Co., Michael Kowalski, *"the factory was built with a focus on the future"*. Laurelton Diamonds employs 140 people in Botswana, which is less than half of full capacity. *"There is ample opportunity for growth"*, he said, *"and the company is eager to continue investing in training to develop the skills of its workforce"*.

Giving a keynote address, President Seretse Khama Ian Khama said, *"like the rest of the world, Botswana has traversed turbulent times and faced challenges in the diamond industry hence the government is keeping a close eye on developments such as these.*

*We rely on our diamonds for the up-liftment of our citizens but we have not yet maximised our true potential...... it is important to diversify the diamond industry from just producing rough diamonds to selling finished products"*.

The diamonds are now ready to be used in the diamond jewellery manufacturing stage of the diamond pipeline.

10

# Jewellery Manufacturing

The final stage of the diamond pipeline in Botswana is the manufacture and retail of diamond jewellery, locally, regionally and internationally.

The main market for diamond currently is the USA with about 50% of the market share, however the emerging markets of China, India and other countries in Asia, will probably become major diamond consumers within the next five years, dominating the overall demand for diamond jewellery globally.

Beneficiation, a strategy shared by the Government of the Republic of Botswana and De Beers, ensures that a significant portion of Botswana's most important natural resource, diamonds, stays in the country longer, so that value can be added through local companies and local expertise. Botswana's downstream diamond industry currently employs nearly 3,000 people.

# Jewellery Manufacturing

Diamonds look dazzling in many forms and the endless options in shape, size and cut can be overwhelming for the average customer when buying a diamond. Knowing what the common diamond shapes are can help one decide if one has a preference for a certain type. Here is a basic rundown:

## ROUND BRILLIANT DIAMONDS

These are the most common kinds of diamonds, accounting for more than 75 percent of diamonds sold today. The round brilliant diamonds include 58 different facets. Round brilliant diamonds are popular in engagement rings, stud earrings and pendants. Because of the high demand for round brilliant diamonds, they are more expensive than many diamond shapes.

*Cut and polished gems (Steinmetz Botswana, 2011)*

# Jewellery Manufacturing

## OVAL DIAMONDS

Oval diamonds are often cut in the 58-facet brilliant style. Ovals are popular shapes for engagement rings as well as for three-stone anniversary rings. Oval diamonds can have an elongating effect on hands, helping women with short fingers look like they have longer ones. Ovals are popular and command a high price.

## MARQUISE DIAMONDS

Marquise diamonds have an elongated shape with pointed ends inspired by the Frenchman Marquise de Pompadour. The marquise diamond is shaped like a rugby ball and the brilliant variety has 58 facets. Some marquise diamonds have less facets, usually around 56. Marquise diamonds are often used for rings and offer a change from the usual round and oval diamond rings. These diamonds look best on slender fingers.

*Shining Light Awards Models (DTCB, 2011)*

*Diamond Models at the 2007 Shining Light Awards (DTCB, 2007)*

# Jewellery Manufacturing

## PEAR-SHAPED DIAMONDS

Pear-shaped diamonds are a combination of the oval and marquise shapes, offering a teardrop look. Pear shaped diamonds are used in rings and are also popular for pendants and earrings. Pear-shaped diamonds look good on small or average length fingers.

## EMERALD CUT DIAMOND

Diamonds with an emerald cut have a rectangular shape with cut corners. The emerald cut diamond has 58 facets but due to the way the diamond is cut, it reflects less light than other diamonds with the same amount of facets, making it sparkle less. Emerald cut diamonds offer a classic, regal look and are most often used in rings.

*2008-2009 Overall Winning Piece, Shining Lights Awards 'Baobab Magic', designed by Katja Nilsson and sponsored by Steinmetz (DTCB)*

# Jewellery Manufacturing

## DIAMOND JEWELLERY MANUFACTURING IN BOTSWANA

Private collection

In 1990, Private Collections started its hand made, classic diamond jewellery business in Botswana and became the first company to manufacture in Botswana, albeit on a very small, and exclusive basis. In 2011, with the assistance of PEO Holdings, it launched a very special unearthed Botswana Collection featuring jewellery designs inspired by pristine wilderness areas such as the Okavango Delta of Botswana and includes the elephant, papyrus and water lily collections.

BOTSWANA UNEARTHED
DIAMONDS
FINE JEWELLERY

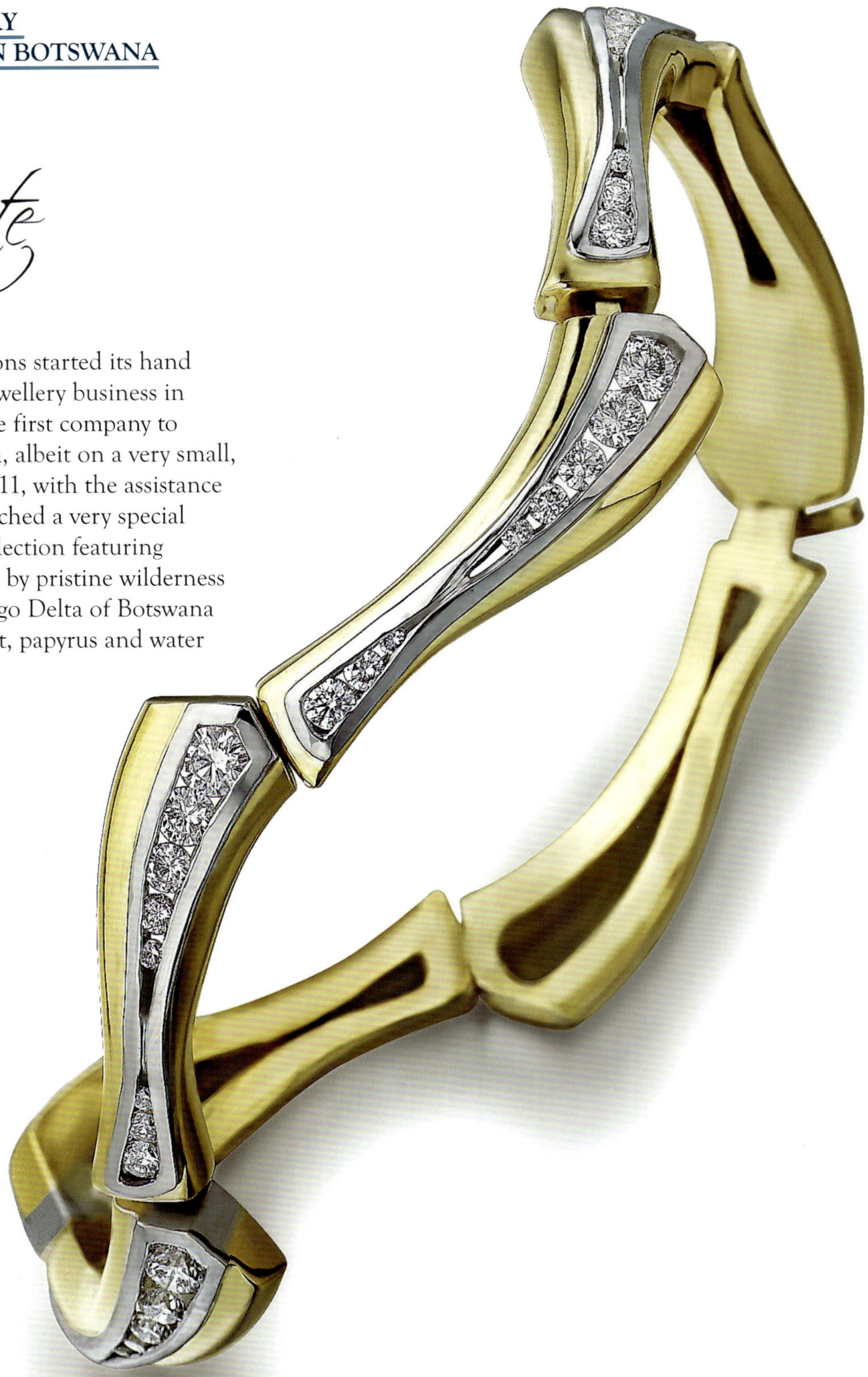

# Jewellery Manufacturing

*President Seretse Khama Ian Khama with Shrenuj MD Shreyas Doshi at the opening of the Shrenuj diamond factory in Gaborone, 7th May, 2010 (Botswana Press Agency)*

The first dedicated large scale diamond jewellery manufacturing plant in Botswana was established by Shrenuj Botswana in the Diamond Technology Park in May 2010. The firm first employed 155 people with plans to grow to 300 employees, mainly making jewellery for the United States market. Botswana has preferential access to the US market under an African trade deal.

*Manufactured in Botswana by Shrenuj (Shrenuj, 2010)*

*2010-2012 Winner - Highly Commended, Shining Lights Awards 'Gold Choker', designed by Tjipo Keaikitse and sponsored by Dalumi Group, Motiganz Diamond Group and Yerushalmi Bros Botswana (DTCB)*

# Jewellery Manufacturing

*Botswana model Kaone Kario wearing one of Steinmetz's luxury diamond creations (Steinmetz, 2008)*

Botswana currently produces rough diamonds worth around US$ 3.5 billion annually. Of this, stones worth about US$ 500 million are earmarked for local processing. Presently, 21 sight holders are engaged in processing diamonds and there is a steady increase in the production of polished diamonds.

The availability of high quality rough diamonds from the local source is the main incentive that diamond manufacturing units get in Botswana. Currently, only a small proportion of the earmarked capacity is being utilised and there is enough room for local units to grow, since Botswana is the largest producer of rough diamonds by value and volume in the world. Shenuj Botswana has sustained its commitment to societal beneficiation by recruiting a third of its employees from the speech and hearing impaired section of the society.

Commenting on this at the launch of the Shrenuj factory, President Khama said, *"I am informed that the Shrenuj Group employs 155 employees, 30 percent of whom are hearing impaired. This answers the Vision 2016 pillar of a Compassionate, Just and Caring Nation and resonates well with our policy of citizen economic empowerment."*

In late 2011, Shrenuj Botswana invested up to US$100 million in a new integrated cutting, polishing and jewellery manufacturing plant to be built in Gaborone over the next two years. The company has acquired 16,000 square meters in the Broadhurst suburb of Gaborone, where the plant will be built. Shrenuj also has plans to open a retail outlet for jewellery in Gaborone.

Botswana told international jewellery makers in 2011 attending the first ever meeting of the International Diamond Manufacturers Association (IDMA) to be held in the country that it was ready to create conditions to entice them to set up shop in the country. Botswana is *"committed to slashing red tape for cutting, polishing and manufacturing jewellery"* said Ponatshego Kedikilwe,

IDMA
INTERNATIONAL DIAMOND MANUFACTURERS ASSOCIATION

Minister of Minerals, Energy and Water Resources at the IDMA meeting. He said, *"Our aspirations are to see as much beneficiation as possible done locally and on a sustainable basis...and we are willing to do what it takes to ensure that an environment that enables beneficiation is created."*

"Beneficiation" has long been part of the Botswana Government's efforts to diversify the economy by encouraging the growth of businesses that add value to diamonds, rather than simply exporting uncut stones.

IDMA was founded and convened for the first time in 1946, in Antwerp. It is committed to fostering and promoting the highest ideals of honesty and best practice principles throughout the diamond industry worldwide, as well as full compliance with all relevant national and international laws.

Towards this end, the organisation has developed and adopted a Code of Conduct, which is binding on all members and works to encourage fair and honourable practices and decent working conditions for those employed in the diamond industry; and to preserve, protect and promote the trust and confidence of consumers in diamonds and diamond products.

IDMA seeks to take a unified leadership role on all issues affecting the industry and convenes biennially at the same time as the World Federation of Diamond Bourses (WFDB) World Diamond Congress. The association has played an active role in establishing the Kimberley Process Certification Scheme and is a co-founder of the World Diamond Council.

# Jewellery Manufacturing

*Pauline Paledi-Mokou, Executive Director*
*Botswana Diamond Manufacturers Association*
*(M. Brook, 2012)*

BOTSWANA
DIAMOND
MANUFACTURERS
ASSOCIATION

According to Pauline Paledi-Mokou, Executive Director of the Botswana Diamond Manufacturers Association (BDMA) *"the BDMA was established in 2007 and consists of 15 out of 16 Diamond Manufacturers in Botswana. The main objective of the organisation is to represent the cutters throughout all facets of Botswana society and government and to assist with government policies such as labour and employment. A lot of investment has been made in the area of training and development, including capacity building of the citizens of Botswana in the use of latest technologies in the diamond manufacturing industry".*

## SHINING LIGHT DIAMOND JEWELLERY DESIGN AWARDS

In 2009, the inaugural, biennial Diamond Trading Company Botswana Shining Light Diamond Jewellery Design Awards showcased the creativity and talent of designers from Botswana. The objective of these diamond-studded awards is to celebrate the beauty and purpose of diamonds through jewellery design and to provide a platform for local designers to showcase their work internationally. This prestigious competition which seeks to create exquisite diamond jewellery as well as unearth the local jewellery design talent was first launched in South Africa in 1996, and will be held biennially.

According to Varda Shine, CEO of DTC, *"the identification and development of new design talent is crucial to the success of the diamond industry and programmes such as the Shining Light Diamond Design Awards are the ideal vehicle for unlocking the wealth of talent in southern Africa. As diamonds compete with other luxury goods for share of consumers' spending, fresh, exciting and innovative designs are required and this is exactly what the Shining Light Awards provide".*

Most of the original 16 DTCB sight holders have sponsored a very special and unique piece of jewellery for the 2009 inaugural Diamond Trading Company Botswana Shining Light Diamond Jewellery Design Awards which showcased the creativity and talent of designers from Botswana.

*From right, Brian McDonald, MD DTCB from 2007 - 2011, Varda Shine, CEO DTC from 2005 - present and Boikobo Paya, Permanent Secretary, Botswana Ministry of Minerals, Energy and Water Resources attending the 2011 shining lights awards in Gaborone (DTCB, 2011)*

In line with the DTC Botswana's Vision of enhancing the emotional value, desirability and integrity of diamonds, the biennial competition encourages designers to create work that is timeless, elegant, mythical and contemporary.

At the 2010-2012 award ceremony, Brian McDonald, the then Managing Director of DTCB said "*The heightened level of enthusiasm from local designers is all the more reason why the Shining Light competition continues to brim with colour, variety and passion. The design submissions received are of a very high standard and are a true reflection of the designers' resolve to compete at international level. The competition has generated a lot of media and public attention and has shown tremendous growth since inception in Botswana two years ago. The competition is also a realisation of a goal envisaged by the Millennium Development Goals – developing a global partnership for development.*

*Part of the competition's success is a result of international companies who share our corporate values and have displayed unyielding faith and generosity. This partnership has a trove of rewards for all participants and we implore Batswana to make full use of this golden opportunity by fully participating in the Shining Light competition*".

Honourable Ponatshego Kedikilwe Minister of Minerals, Energy and Water Resources added, "*It is gratifying to note that these awards are at the heart of our key strategic initiatives of citizen economic empowerment and development of diamond jewellery manufacturing in Botswana. The awards indeed recognize individual achievement and serve as an important motivational tool for jewellery designers to excel in their artistic pursuits. This year's collection bears testimony to DTC Botswana's beneficiation drive of adding real value to the country's downstream diamond industry thereby ensuring long-term sustainability of the diamond industry in Botswana*".

# Jewellery Manufacturing

## DTC Botswana Shining Light Awards

### 2010-2012

Overall Winning Piece

**David Moatisi**
sponsored by Eurostar

ZEBRA NECKLACE
18ct white gold neck piece set with 2,848 round brilliant black and white diamonds with a total weight of 73.17cts

*"The love of a country begins an attachment that harnesses actions of both honour and dishonour but never indifference."*

### 2010-2012

Winning Piece

**Ditshupo Mogapi**
sponsored by Shrenuj

KUDU HORNS
18ct white and yellow gold neckpiece set with 344 round brilliant cut diamonds with a total weight of 16.16cts

*"When my tongue tasted dust and the sunburnt veld cracked beneath my feet, our most loving Mama obliged Africa and swept our plains with rain"*

### 2008-2009

Overall Winning Piece

**Katja Nilsson**
sponsored by Steinmetz

BAOBAB NECKLACE
18ct White and Red Gold, set with 674 round brilliant cut Diamonds, weighing 24.95ct.

*"Inspired by Kubu Island in the Makgadikgadi Pans, Botswana.The Majestic, Ancient Baobab Trees stand like wise old people with outstretched hands awaiting the rains. And at night ... the Stars twinkle through their branches like Diamonds".*

### 2010-2012

Winning Piece - Highly Commended

**Tjipo Keaikitse**
sponsored by Dalumi Group, Motiganz Diamond Group and Yerushalmi Bros Botswana

GOLD CHOKER
18ct white gold choker set with 3,425 round brilliant cut diamonds weighing a total of 17.169 cts.

*"Faith, hope, despair, belief; like pieces of a jigsaw, each piece comes together to make sense of the puzzle that is Life".*

## DTC Botswana Shining Light Awards

### 2010-2012
Winning Piece

**Buhle Mguni**
sponsored by Mohit Diamonds Private Limited

ARM BRACELET
18ct white gold & rhodium plated Sun Cuff set with 887 round brilliant cut diamonds with a total weight of 17.43cts

*"The warmth from the sun awakens the earth's sleep, heaven smiles and love no longer weeps"*

### 2008-2009
Winning Piece

**David Atho Moatisi**
sponsored by Shrenuj

MASK
18ct yellow gold mask set with 1,172 round brilliant cut diamonds weighing 15.23cts and 31 Canary coloured diamonds weighing a total of .62cts

*"Each day I wake-up I am bathed in the warmth of the sunrise, and I am able to face the promise of another day with renewed challenge and vigour. In the same way that the sun invigorates my body, my soul is invigorated by the challenge of new opportunities".*

### 2010-2012
Overall Winning Piece

**Grace Hanson**
sponsored by Teemane Manufacturing

PONYTAIL BRACELET
18ct white gold hair piece set with 80 round brilliant cut diamonds weighing 22.86cts.

*"The beauty of youth reflected in confidence manifested in laughter; the beauty of age suggested by peace exposed by wisdom; physical beauty imitated - inner beauty ignored – why is this so"*

# Jewellery Manufacturing

*Botswana's Diamonds at work and at home (M. Brook, 2011)*

An incredible thirty pieces of classical and beautiful jewellery, containing no fewer than 25,000 separate diamonds, were on display at Botswana's 2010-2012 Shining Light Awards judging held at the DTCB Headquarters. The Diamond Technology Park's slogan *"Botswana's Diamonds @ work @home"* is fitting when one considers that diamonds are not just mined in Botswana, but also sorted, valued, cut, polished and placed into sparkling jewellery, before being sold locally, regionally and globally.

One hundred and six (106) entries were received in the 2010 competition from professional designers, students and apprentice jewellery designers from across Botswana. With international interest and awareness in this prestigious competition on the increase, the

*Sheila Khama, first female Chief Executive of De Beers Botswana between July 2005 and March 2010 (Debswana, 2010)*

*Diamond Jewellery Manufacturing in Botswana (Shrenuj, 2011)*

2010 Judging Panel comprised international judges including supermodel Alek Wek (Sudan), international jewellery designer, Karen Lee (China), and Art Consultant Yoshiko Ikoma (Japan). Local judges comprised Jacob Thamage (Coordinator of the Diamond Hub); Mildred Boduwe (founding Principal of the Oodi College of Applied Arts and Technology); Philisani Amos (Lecturer, Limkokwing University of

*Motswana jewellery designer, David Moatisi, with his award winning piece at the 2011 Shining Light Awards in Gaborone. (DTCB)*

*Kim Kardashian wearing De Beers Forevermark Jewellery (De Beers, 2009)*

Creative Technology); Sheila Khama (former Chief Executive of De Beers Holdings Botswana); and Brian McDonald (the then DTC Botswana Managing Director).

David Moatisi's zebra necklace was the winning design; an 18ct white gold neck piece set with no fewer than 2,848 round brilliant black and white diamonds with a total weight of 73.17 cts.

De Beers has also introduced it's own diamond jewellery brand, "Forevermark diamonds", and sells mostly to markets in China, Hong Kong, India and Japan. According to the company, Forevermark diamonds, *"are natural, untreated, responsibly sourced, and cut and polished by a specially selected diamantaire."* Forevermark diamonds have an icon and identification number inscribed on the table facet of the diamond. The inscription is about 0.05 μm deep.

# Jewellery Manufacturing

*0.01 ct of 60 pieces with white and yellow gold (Schonberg Jewellers, Gaborone, 2011)*

There are also a small, but increasing, number of shop based jewellery manufacturing workshops in Botswana, mostly located in the capital city, Gaborone. One such is Schonberg jewellers who have produced some unique necklace pieces of Botswana diamond encrusted agate.

As for public training and education, the new College of Applied Arts and Technology, which opened in Oodi in 2012, offers, for the very first time in Botswana, a certificate, advanced certificate and Diploma in Jewellery Design and Manufacturing. The course lasts between 12-18 months, and 32 students are expected to enrol in 2012.

The opportunities for further diamond jewellery manufacturers to establish in Botswana are tremendous and many more companies are expected in the country in the next few years

*Botswana Agate and Diamond pendant with 9kt white gold (Schonberg Jewellers, 2011)*

*Oodi College of Applied Arts and Technology - home of the first Diamond Jewellery Diploma in Botswana (M. Brook, 2011)*

to take advantage of the favourable business environment and the guaranteed supply of cut and polished diamonds from the established sightholders.

The final stage of the diamond pipeline is, in Botswana's case, to expand the beneficiation process so that Botswana becomes a recognised diamond hub for all the diamond pipeline activities from exploration, to mining, sorting, valuing, cutting, polishing, jewellery manufacturing and finally sales.

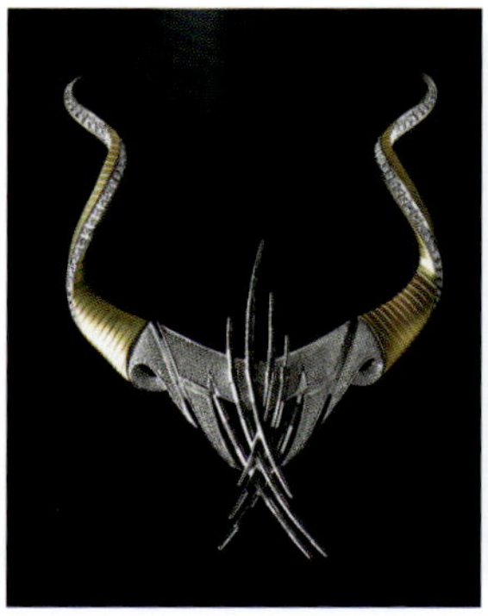

*Kudu Horns Necklace Shining Lights Awards, 2011 (Shrenuj, 2011)*

*The Mask, Shining Lights Awards Winner, 2009 (Shrenuj, 2011)*

11

# Beneficiation

In the diamond industry the term beneficiation refers to downstream activities in a producer country such as Botswana that adds value to locally mined rough diamonds and creates local employment.

It includes the process of sorting and valuing rough diamonds, their subsequent cutting and polishing, and the manufacture of diamond jewellery. This strategy of turning natural resources into shared national wealth is anticipated to stimulate the development of a new industry in Botswana and create many more jobs in the country. The jobs, primarily in diamond cutting, polishing, jewellery manufacture, sales and marketing, will account for a 30% increase in diamond industry jobs in Botswana and result in a 10% increase in the country's total number of manufacturing jobs. In addition, job creation in secondary business sectors such as banking, security and information technology will support Botswana's economic vision to generate jobs and drive maximum value from the country's diamond resources.

# Beneficiation

*Gareth Penny, MD De Beers, 2006 to 2010. (Debswana) and De Beers CEO, Phillipe Mellier (DTCB, 2011)*

*Examining small rough stones at Orapa House, Gaborone (BDVC, 2007)*

Gareth Penny, the then Group Managing Director, De Beers, speaking at the Antwerp Diamond Conference, 15th October 2007, said "*For The African diamond producing countries, beneficiation is not optional, not a passing whim motivated by political correctness, but an imperative, an absolutely essential and critical part of the macroeconomic policy designed to uplift their economies to provide education, jobs and healthcare for their people, and to make poverty history.*"

The two stages of the diamond pipeline in Botswana which are yet to be fully developed are diamond cutting and polishing and diamond jewellery manufacturing. The development of the Diamond Technology Park is a demonstration that Government and the private sector are seriously committed to strengthening activities in these two stages of the diamond pipeline.

According to Chaim Even-Zohar (2007), the average cost of just sorting and valuing Botswana's diamonds is about US$0.50 /carat, thus, with an average annual production of 30 million carats, it will provide Botswana with some US$15 million in added value for the national economy. Signficant added value will further be derived from cutting, polishing and manufacturing diamond jewellery.

The inaugeral 2011 Botswana meeting of the presidents of organisations affiliated to the International Diamond Manufacturing Association (IDMA) put the spotlight back on the issue of beneficiation. This came after the financial crisis and economic slowdown during 2008 to 2010 had diverted attention from the efforts of African diamond producer states to retain a higher percentage of the added value provided by their natural resources.

# Beneficiation

*Orapa rough diamond, which was cut and polished by Steinmetz and presented to Miss Universe Botswana, Tirelo Ramasedi, to take to Las Vegas for the nation's representation at the Miss Universe Beauty Pageant. (Steinmetz, 2010)*

While beneficiation was largely a South African initiative, launched during the presidency of Thabo Mbeki, it has been adopted with vigour by that country's northern neighbours, and in particular Botswana, which has insisted on diamond manufacturers establishing local cutting plants in order to obtain direct supply from the country's rough output.

With Botswana being the world's largest producer of diamonds, Dr. Ponatshego Kedikilwe said *"we are now ready to create conditions to entice the international diamond jewellery industry to set up shop in the country. Our aspirations are to see as much beneficiation as possible done locally and on a sustainable basis, we are willing to do what it takes to ensure that an environment that enables beneficiation is created."*

Also speaking at the meeting, IDMA President Moti Ganz made it clear that the power in the diamond industry was now with the producer countries: *"We have come here because we are looking to the future. We all say that these days we have to think outside the box. We must realise that, today, the producers are the leading voice in the diamond sector and they are the ones who must lead the industry into the future"*, he said.

*Head of Mining – Debswana Diamond Company, Lenayang Dimbungu (M. Brook, 2012)*

*"We want to understand, as stakeholders in the industry, what problems we face and how we can come up with solutions for these challenges,"* said Botswana Diamond Manufacturers Association (BDMA) President Mervin Lifshitz. *"How do other stakeholders, both in the country and outside, join together to contribute to the development of the country as well as our businesses?"*

On 25th September, 2011, Pauline Paledi-Mokou, a Motswana, was appointed as the BDMA Executive Director, replacing Mervin Lifshitz.

The beneficiation drive brought about a massive restructuring of De Beers' rough diamond distribution machine, leading to the establishment of DTC subsidiaries in South Africa, Botswana and Namibia. Prior to this, all diamonds were channelled through the DTC headquarters in London.

According to the Botswana Government, in November, 2011, Botswana's diamond beneficiation industry employed almost 3,262 (88% of whom are Batswana), as well as creating work in related sectors, such as banking, security and transport. Recently, Suashish Diamonds, an Indian diamond polisher and retailer announced, that it will also set up a jewellery manufacturing plant in Botswana.

The Indian Diamond Institute (IDI) will have a presence in Botswana during 2012. A reputed teaching institution, IDI has been chosen by the Indian government to help set up the India-Africa diamond institute at the Oodi College of Applied Arts and Technology, some 15km north of the capital city, Gaborone where Botswana's first diploma course in Jewellery Design and Manufacturing will be offered.

For Botswana, the proposed India-Africa Diamond Institute is a timely development as the country moves beyond diamond cutting and polishing towards greater beneficiation.

The Botswana National Human Resource Development Strategy (NHRDS), approved in 2009, identified a number of sectors which are critical players in the strategy to increase economic growth and create employment in the country. The vision of the NHRDS is the development of an economy in which *'the quality, productivity and motivation of its people will be Botswana's single greatest and valuable resource'* (Vision 2016).

# Beneficiation

*Mine conveyors, Jwaneng Mine (Debswana)*

In 2008, the Botswana Government clustered a number of major development projects into six Hubs to attract internal and external investment:

- **A Diamond Hub** to facilitate beneficiation and promote Botswana as one of the world's major diamond trading centers.
- **A Transport Hub** to expand and promote synergy in the air, road, and rail sectors.
- **An Agricultural Hub** to serve as a catalyst for the commercialisation and diversification of the sector, as well as to improve food security.
- **A Health Hub** to make Botswana a centre of excellence in the provision of healthcare services for medical tourists as well as local patients.
- **An Education Hub** to enable Botswana to have a competitive edge both regionally and globally through the promotion of education, training and academic research
- **An Innovation Hub**, which will serve as a commercial nexus for applied research by catering for ICT and Research and Development companies.

A strategy for the development of diamond cutting, polishing and jewellery making skills is also being put in place. This strategy supports the introduction of institution based diamond cutting and polishing training in Botswana. Diamond cutting, polishing and jewellery is a fast growing industry of great strategic importance to Botswana. Employment in this sector has trebled in the last four years, and is expected to reach 6000 employees by 2016.

*Dr Akolang Tombale, Head of the Botswana Diamond Hub from 2008 - 2010 (Debswana, 2010)*

*Jacob Thamage, Head of the Botswana Diamond Hub 2010 - present (M. Brook, 2012)*

The Botswana Diamond Manufacturing Association envisages an ideal scenario in which there would be a very close working relationship between employers and an institution that provides training in basic skills, as well as in specialist areas, as indicated by the industry. In this scenario individual students would be 'groomed' in response to specific company demands; attached to that company during training; and 'offered' as an employee to the company after graduation. Furthermore, there is a desire to form win-win arrangements between companies, which facilitate the secondment of employees between companies, in order to provide employees with broad knowledge of the industry, and to share knowledge and skills among companies.

A number of neighbouring SADC countries have been very successful over the last 40 years or so in establishing a downstream diamond business. For example, Mauritius, which does not produce rough diamonds, employs 4,200 people in 420 domestic jewellery manufacturing units whose combined production amounts to no less than 9% of manufactured exports.

Therefore, opportunities for beneficiation in Botswana are enormous!

Dr Akolang Tombale, former head of the Botswana Diamond Hub said in Gaborone in 2009, *"Our wish as a country to become a big player in the Downstream Diamond Industry is not born out of sentiment of being the world's leading producer but from genuinely believing that there is an opportunity to be taken as the time is right to take the plunge. However, we believe that that should not be a solitary effort but should be shared with entrepreneurs' out there who believe in the dream".*

According to Jacob Thamage, the current Head of the Botswana Diamond Hub, Government has two main challenges;

- The need to embrace free open market principles;
- To provide an enabling environment within which private business can most efficiently organize production to profitably meet demand and services both at home and abroad.

# Beneficiation

The Diamond Hub, established in 2008 and housed within the Diamond Technology Park, is providing support for the development of diamond beneficiation and related processing activities, along with the Diamond Office and Botswana Diamond Manufacturing Association. Its objectives are to provide an environment for efficient diamond sector operation and make the sector economically attractive and also to start critical activities of the diamond value chain in Botswana to transform the country into a fully integrated Diamond Centre. The following activities are ongoing: (Botswana Diamond Hub, undated):

*Diamonds mined, polished and jewelled in Botswana (Private Collection 2012)*

1. To improve the commercial and economic viability of the cutting and manufacturing component of downstream diamond processes in Botswana;
2. To establish the Botswana Diamond Exchange to sell both rough and polished diamonds locally;
3. To facilitate the development of downstream diamond support businesses such as financing, technology and security;
4. To develop a niche market for diamond manufacturing in Botswana;
5. To facilitate the development of diamond tourism which may include mining areas' tours; diamond factories' tours, diamond museums and diamond jewellery shopping;
6. To establish a Botswana Diamond Research and Training Centre to train both diamond cutters and young entrepreneurs.

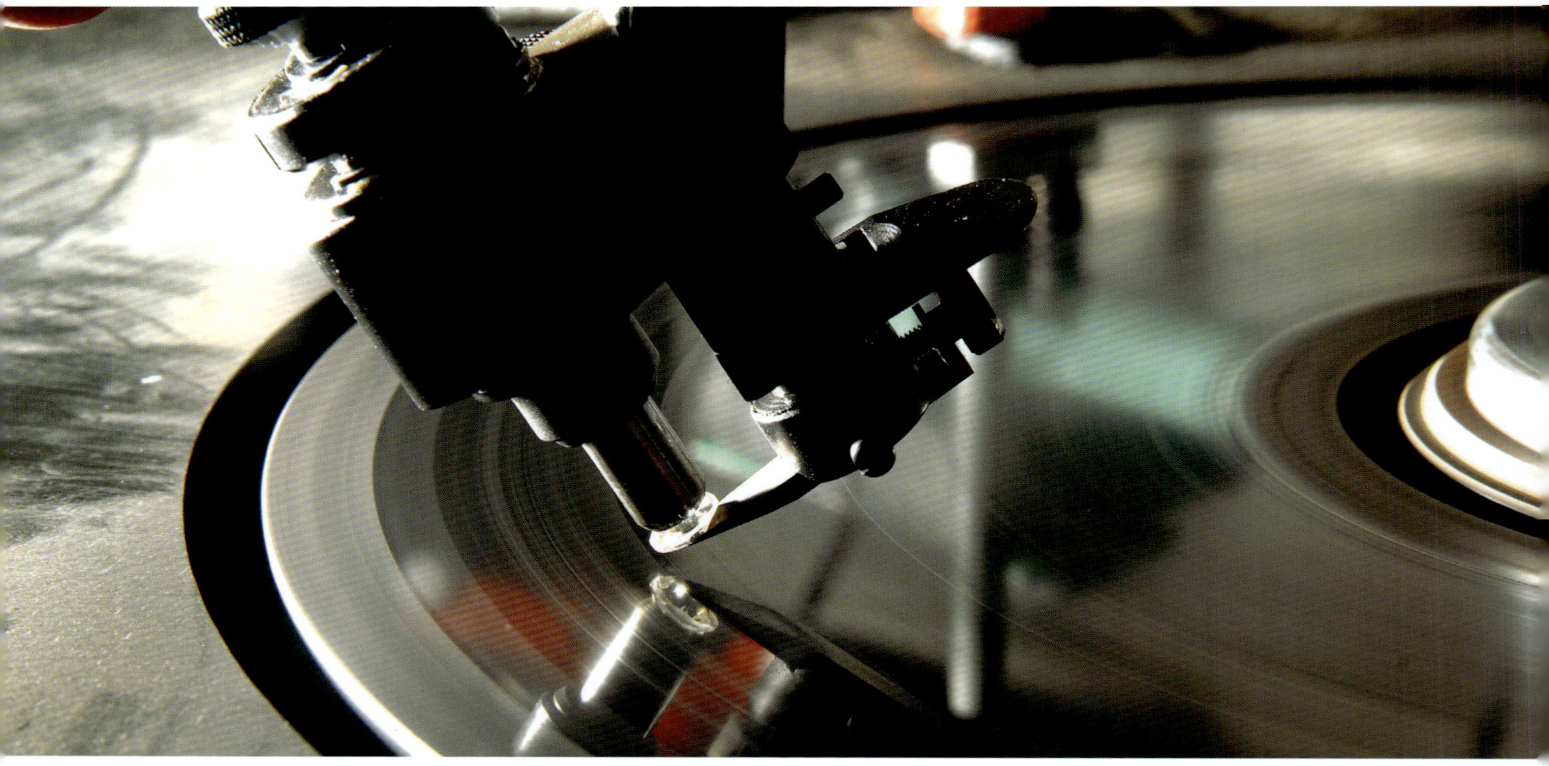

*Diamond polishing (Steinmetz, 2010)*

According to Thamage, examples of creating an enabling environment for the Botswana diamond industry include:

- Increased availability of rough daimonds;
  - Local sales from new mines
  - Minimum supply by DTCB
- Expatriates working in the cutting and polishing industry exempted from full requirements of residence and work Permits
- diamond cutting and polishing companies exempted from paying training levy
- Cutting and polishing factories designated bonded facilities for VAT purposes (improves cash flow)
- The introduction of special economic zones

The journey of Botswana's Diamonds in the diamond pipeline is shown below:

# The Journey of Botswana Diamonds

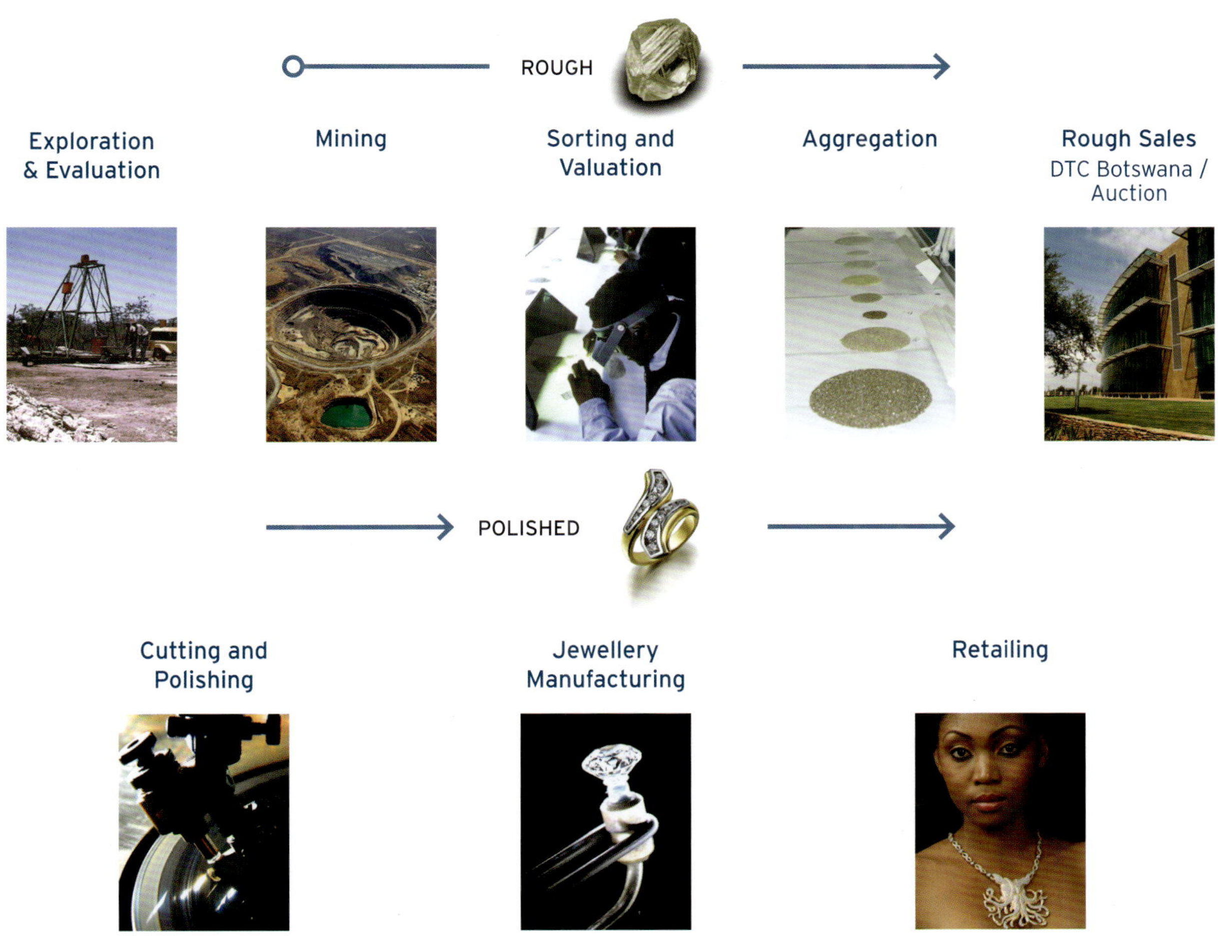

# Beneficiation

Botswana has a strong desire to strengthen its participation in the five most downstream sectors shown above.

The Botswana Government has also financed the construction of a Secure Transfer Facility (STF) at Sir Seretse Khama International Airport. Since diamond activities are usually developed around a central point, the area around the DTCB building and Debswana Corporate Center are the first phase of what would become a Diamond District, since a cluster of diamond activities (located mostly within the Diamond Technology Park) are concentrated in the area with direct, secure road linkage to the airport via the STF.

*Blackie Marole, Debswana Managing Director, 2004 - 2010 and Jim Gowans, Managing Director Debswana, 2011 - Present (Debswana)*

In 2011, DTCB shareholders, De Beers and the Botswana Government, signed a new 10-year sales agreement which also provides for the relocation of DTC from London to Gaborone, a move that is expected to be completed by 2013. De Beers Group CEO, Phillipe Mellier, said in Gaborone in 2011 at the De Beers - Botswana Government 10-year agreement ceremony, *"the relocation is a big corporate coup for Botswana as it will further entrench and consolidate the country as a world leader in the global diamond industry"*. He further said *"the fact that De Beers will become much more visible in Botswana as a result of the relocation will bring with it many spinoffs to the economy of Botswana, including retail, banking, security, aviation, housing, schools and many other sectors"*.

The re-location of DTC International from London to Botswana means that aggregation is due to begin in 2013 at DTCB's centre of excellence in Gaborone. *"We will have a carbon copy in Gaborone of what happens in London,"* said DTC Botswana Sales Head Marcus ter Haar.

In March 2012, the De Beers Aggregation Company was formed in Botswana. As well as its aggregation function, the sales and trading activities of DTC London will be relocated to Botswana. Eventually, the London "sight" will disappear and be transfered to Gaborone by the end of 2013.

A press release by the Government of the Republic of Botswana on the agreement, signed 16th September, 2011, said that this *"Partnership solidified a long-term agreement to make Botswana a leading diamond hub.*

*The Government of the Republic of Botswana and the De Beers Group today announced that agreement has been reached on a new 10 year contract for the sorting, valuing and sales of Debswana's diamond production. As part of the milestone agreement, De Beers will transfer its London-based rough diamond sales activity to Botswana, underpinning the long-term future of the partnership and transforming Botswana into one of the world's leading diamond trading and manufacturing hubs.*

*The agreement, which commences from 1 January 2011, is the longest sales contract ever agreed between the two partners. It will see the Diamond Trading Company (DTC) relocate its Sights and sales operations - including professionals, skills, equipment and technology - from London to Gaborone by the end of 2013. From its new base in Botswana, the DTC will aggregate production from De Beers' mines, and its joint venture operations worldwide, and sell to international Sightholders. Diamond Trading Company Botswana (DTCB) will continue to sort and value Debswana's production before selling it on to the DTC. The DTC will continue to support the domestic cutting and polishing industry in Botswana, and will increase its commitment by making more diamonds available for manufacturing businesses operating in the country".*

*Botswana's Minister of Minerals, Energy and Water Resources, Dr. Ponatshego Kedikilwe and Nicky Oppenheimer, De Beers Chairman at the $16^{th}$ September 2011 signing of the agreement (DTCB)*

# Beneficiation

*Botswana's Minister of Minerals, Energy, and Water Resources, Dr. Ponatshego Kedikilwe (DTCB, 2011)*

*Nicky Oppenheimer, Chairman of De Beers 1997 to 2011 (Debswana, 2011)*

In addition, the agreement provides for an independent sales outlet for the Botswana Government, which began in 2011 at 10 percent of Debswana's run of mine production and rising to 15 percent over a five-year period. The Minister of Minerals, Energy and Water Resources, Dr. Ponatshego Kedikilwe, hailed the agreement as a major step in securing the future of Botswana's economy, saying, *"Botswana and De Beers have today renewed one of the most successful public/private partnerships in the world. This agreement, and the tangible outcomes it will deliver, will enable Botswana to achieve its aspiration to be a major diamond centre engaged in all aspects of the diamond business.*

*The agreement gives us direct access to market, which will be helpful for the successful development of the downstream diamond industry in Botswana. The agreement will also bring more employment and other opportunities to Botswana. Our agreement is only the first step - much work needs to be done during the next several years to make this transformation a success. As a major player in the diamond trading industry we look forward to welcoming the global diamond industry to Botswana. We are invigorated by the challenge and excited by the possibilities."*

Nicky Oppenheimer, Chairman of De Beers, commented, *"With this agreement, De Beers cements its commitment to being the driving force behind Botswana's emergence as a world-leading diamond trading and manufacturing centre. The transformation that this agreement sets out is testament to our belief in the future of Botswana, and is a clear acknowledgment that De Beers' success is only possible with a profound understanding that the aspirations of our partners must be at the core of our business. I am proud of this moment and excited by what we will build together."*

Mr. Oppenheimer added that, as a result of the Agreement, *"Botswana has preserved and enhanced a highly successful route to market, focused on maximising the value of her natural resource, and De Beers has secured long-term and uninterrupted access to the largest supply of diamonds in the world.*

*Varda Shine, CEO, DTC from 2005 - present (DTC, 2011)*

*Eric Molale, Permanent Secretary, office of the Botswana President and Deputy Chairperson of Debswana, 2011 (Debswana)*

*Neo Moroka, appointed resident Director and Chief Executive of De Beers Botswana Holdings in February, 2010 (Debswana)*

*This is a once in a generation opportunity for Botswana and De Beers to shape the future of the diamond industry, and we look forward to working with our partners to provide our Sightholders with the continuity and quality of supply they expect while pushing beneficiation to greater levels than ever seen before."*

Varda Shine, CEO of the Diamond Trading Company, the sales and marketing arm of the De Beers Group said that "*The DTC is committed to beneficiation and, as part of the new contract period, more De Beers diamonds than ever before will be made available to a larger pool of sightholders in Botswana and Namibia,*" she explained.

According to Eric Molale, chairperson of Debswana 2012-2013 and Botswana's lead negotiator with De Beers in the above agreement, "*We as Botswana, will have to reciprocate. We also have to give back to the world. Key amongst those that we have to do in that regard is productivity....... this agreement will bring a positive impact and further consolidate our existing joint ventures as well as nurturing the Botswana diamond industry to grow into a force to be reckoned with worldwide.*"

Neo Moroka, appointed resident Director and Chief Executive of De Beers Botswana in February, 2010 said, "*The new agreement transforms Botswana into one of the leading countries in diamond trading and manufacturing. We are happy that this Agreement underpins the future of the partnership between Botswana and De Beers following assumptions formulated in the past as regards the same partnership.*"

Moroka observes that De Beers would not ignore the fact that the success of the producer countries' economies are at the core of the business.

# Beneficiation

*Linah Mohohlo, Governor, Bank of Botswana and Dr. Athaliah Molokomme, Attorney General of Botswana, Debswana Board Members (Debswana, 2011)*

Spadework has begun on construction of a trading facility at Diamond Technology Park (DTP) in Gaborone, marking the first stage of what will become a platform for independent tenders for local, regional and international diamond producers and buyers trading in Botswana. It will be Africa's very first Diamond Auction Centre, open to a transparent tender process. The Government diamond tender house is expected to be operational by the second half of 2012. In March 2012, the Botswana State Diamond Company, called the Okavango Diamond Trading Company was formed.

Infrastructure at the three-storey turnkey project that is being built on 1,410 square metres will include sorting rooms, viewing rooms, management offices, a laboratory, a control room, a café, technology hardware, diamond equipment and internal security systems that include CCTV and access control.

Says Rutang Moses, the Chief Executive Officer of DTP: *"Establishment of the trading facility marks the next step towards Botswana becoming an international trading centre for rough and polished diamonds. DTP anticipates this project will encourage diamond producers to sell their production in Botswana, offering diamonds to the newly established cutting industry, thereby underpinning viability of the trading facility."*

However, before Botswana can succeed in bringing regional and international producers to trade their production in Gaborone, the Government owned Okavango Diamond Trading Company's new diamond auction centre needs to be completed in 2012 and will be the platform for the trading of rough diamond production, unlocked from the acquisition of the 10% run-of mine independent marketing window that the Botswana Government has now negotiated with De Beers.

At present, the existing 16 diamond polishing cutting factories (which are to be increased to 21) need about $800 million worth of supplies annually.

When the window opens, Botswana will process about $1.2 billion locally. As part of efforts to turn Gaborone into a world recognised diamond centre, the government has also given the green light to Firestone and Lucara to sell part of their production from B/K11 and A/K6 mines by open tender. Firestone has held four tenders since December 2010.

*Dr. Ponatshego Kedikilwe, Botswana's Minister of Minerals, Energy and Water Resources with James Kirby, Jwaneng Cut 8 Mining Lead from 2009 - 2012 (Debswana, 2007)*

Breaking the ground for the new diamond trading facility, Diamond Hub Coordinator, Jacob Thamage, commented: *"I'm extremely excited by the start of the development of Phase Two of the DTP. It couldn't have come at a better time."*

For over 100 years diamonds have left Africa to be sorted, valued and sold in London. The announcement of the new 10-year Botswana Government – De Beers deal means that Botswana diamonds will stay in Botswana for longer to have more value added and to enable Botswana to derive more from what is essentially a finite natural resource.

DTC Botswana will do the aggregation - putting like diamonds together from various mines and countries - which is currently done in the UK for all of De Beers diamonds, including those from South Africa, Namibia and Canada, in facilities that will replicate those in London.

The scene is at last now set for Botswana to potentially become one of the most important Diamond Centers of the world!!!

# Appendix

## Botswana's Kimberlites

| Kimberlite Name | Latitude | Longitude | Date Discovered | Prospecting Company |
|---|---|---|---|---|
| MOCHUDI-01 | -24.388839 | 26.397837 | 1966 | DEBOT |
| MOCHUDI-02 | -24.331871 | 26.380531 | 1966 | DEBOT |
| LETLHAKANE_DK1 | -21.520283 | 25.689518 | 1967 | DEBOT |
| ORAPA_AK1 | -21.308725 | 25.369808 | 1967 | DEBOT |
| ORAPA_BK6 | -21.333947 | 25.592016 | 1967 | DEBOT |
| ORAPA_BK7 | -21.348666 | 25.580613 | 1967 | DEBOT |
| ORAPA_BK3 | -21.329193 | 25.574515 | 1969 | DEBOT |
| LETLHAKANE_DK2 | -21.524217 | 25.695912 | 1970 | DEBOT |
| ORAPA_AK10 | -21.460271 | 25.482915 | 1970 | DEBOT |
| ORAPA_AK11 | -21.389145 | 25.449208 | 1970 | DEBOT |
| ORAPA_AK12 | -21.322856 | 25.431011 | 1970 | DEBOT |
| ORAPA_AK13 | -21.377851 | 25.439105 | 1970 | DEBOT |
| ORAPA_AK14 | -21.341093 | 25.469113 | 1970 | DEBOT |
| ORAPA_AK2 | -21.313076 | 25.390109 | 1970 | DEBOT |
| ORAPA_AK3 | -21.292024 | 25.361511 | 1970 | DEBOT |
| ORAPA_AK4 | -21.293407 | 25.411508 | 1970 | DEBOT |
| ORAPA_AK5 | -21.322264 | 25.397007 | 1970 | DEBOT |
| ORAPA_AK6 | -21.496466 | 25.470108 | 1970 | DEBOT |
| ORAPA_AK7 | -21.308718 | 25.359803 | 1970 | DEBOT |
| ORAPA_AK8 | -21.405863 | 25.430609 | 1970 | DEBOT |
| ORAPA_AK9 | -21.418647 | 25.445405 | 1970 | DEBOT |
| ORAPA_BK1 | -21.279211 | 25.614512 | 1970 | DEBOT |
| ORAPA_BK10 | -21.403411 | 25.570411 | 1970 | DEBOT |
| ORAPA_BK11 | -21.471189 | 25.511713 | 1970 | DEBOT |
| ORAPA_BK12 | -21.305136 | 25.529210 | 1970 | DEBOT |
| ORAPA_BK13 | -21.309819 | 25.601713 | 1970 | DEBOT |
| ORAPA_BK14 | -21.280098 | 25.750913 | 1970 | DEBOT |
| ORAPA_BK15 | -21.107297 | 25.570108 | 1970 | DEBOT |
| ORAPA_BK16 | -21.407782 | 25.720119 | 1970 | DEBOT |
| ORAPA_BK17 | -21.109016 | 25.570409 | 1970 | DEBOT |
| ORAPA_BK2 | -21.287551 | 25.674212 | 1970 | DEBOT |
| ORAPA_BK4 | -21.308704 | 25.630913 | 1970 | DEBOT |
| ORAPA_BK5 | -21.253079 | 25.547015 | 1970 | DEBOT |
| ORAPA_BK8 | -21.275122 | 25.589811 | 1970 | DEBOT |
| ORAPA_BK9 | -21.312608 | 25.536713 | 1970 | DEBOT |
| JWANENG-01 | -24.648433 | 24.808384 | 1973 | DEBOT |
| JWANENG-03 | -24.641423 | 24.791983 | 1973 | DEBOT |

| Kimberlite Name | Latitude | Longitude | Date Discovered | Prospecting Company |
|---|---|---|---|---|
| JWANENG-04 | -24.642861 | 24.671979 | 1973 | DEBOT |
| JWANENG-05 | -24.618446 | 24.782581 | 1973 | DEBOT |
| JWANENG-06 | -24.614229 | 24.747888 | 1973 | DEBOT |
| JWANENG-07 | -24.548928 | 24.769781 | 1973 | DEBOT |
| JWANENG-08 | -24.653143 | 24.580083 | 1973 | DEBOT |
| JWANENG-09 | -24.627523 | 24.835591 | 1973 | DEBOT |
| KGALAGADI-01 | -24.941439 | 22.126109 | 1975 | DEBOT |
| KGALAGADI-02 | -24.899561 | 22.020503 | 1975 | DEBOT |
| KGALAGADI-03 | -24.537097 | 22.585523 | 1975 | DEBOT |
| KGALAGADI-04 | -24.584824 | 22.614723 | 1975 | DEBOT |
| KGALAGADI-05 | -24.986724 | 22.092203 | 1975 | DEBOT |
| KGALAGADI-09 | -24.591725 | 22.616424 | 1975 | DEBOT |
| KGALAGADI-10 | -24.904216 | 22.223011 | 1975 | DEBOT |
| KGALAGADI-12 | -24.541284 | 22.586923 | 1975 | DEBOT |
| KOLONKWANENG-01 | -26.630488 | 22.019704 | 1975 | DEBOT |
| KOLONKWANENG-02 | -26.082465 | 21.949102 | 1975 | DEBOT |
| MALAGONG_SILLS | -26.783829 | 21.532981 | 1975 | DEBOT |
| MIDDLEPITS-01 | -26.670501 | 21.397984 | 1975 | DEBOT |
| MIDDLEPITS-02 | -26.675488 | 21.407982 | 1975 | DEBOT |
| MIDDLEPITS-03 | -26.667222 | 21.412986 | 1975 | DEBOT |
| MIDDLEPITS-04 | -26.677253 | 21.397984 | 1975 | DEBOT |
| MIDDLEPITS-05 | -26.675556 | 21.384679 | 1975 | DEBOT |
| MIDDLEPITS-06 | -26.715525 | 21.434684 | 1975 | DEBOT |
| MIDDLEPITS-07 | -26.718772 | 21.427988 | 1975 | DEBOT |
| TSABONG_T19 | -26.071354 | 22.158003 | 1975 | FALCONBRIDGE |
| JWANENG-02 | -24.523976 | 24.701183 | 1978 | DEBOT |
| KANG_KE1 | -24.504869 | 22.949730 | 1978 | FALCONBRIDGE |
| KANG_KN19 | -24.267398 | 22.903935 | 1978 | FALCONBRIDGE |
| KANG_KN4 | -24.308941 | 22.899737 | 1978 | FALCONBRIDGE |
| KANG_KN49 | -24.167398 | 22.733032 | 1978 | FALCONBRIDGE |
| KANG_KN63 | -24.184014 | 22.737225 | 1978 | FALCONBRIDGE |
| KANG_KN65/73 | -24.204920 | 22.708024 | 1978 | FALCONBRIDGE |
| KANG_KN70 | -24.150705 | 22.716423 | 1978 | FALCONBRIDGE |
| KANG_KN8 | -24.267349 | 22.862235 | 1978 | FALCONBRIDGE |
| KANG_KN90 | -24.267374 | 22.916437 | 1978 | FALCONBRIDGE |
| KANG_KS10 | -24.467378 | 22.824732 | 1978 | FALCONBRIDGE |
| KANG_KS11 | -24.450649 | 22.816430 | 1978 | FALCONBRIDGE |

# Appendix

| Kimberlite Name | Latitude | Longitude | Date Discovered | Prospecting Company |
|---|---|---|---|---|
| KANG_KS13 | -24.433942 | 22.783028 | 1978 | FALCONBRIDGE |
| KANG_KS27 | -24.521432 | 22.833035 | 1978 | FALCONBRIDGE |
| KANG_KS28 | -24.533983 | 22.866426 | 1978 | FALCONBRIDGE |
| KANG_KS31 | -24.467346 | 22.849734 | 1978 | FALCONBRIDGE |
| KANG_KS34 | -24.446437 | 22.791426 | 1978 | FALCONBRIDGE |
| KANG_KS37 | -24.483998 | 22.758030 | 1978 | FALCONBRIDGE |
| KANG_KS40 | -24.450657 | 22.833026 | 1978 | FALCONBRIDGE |
| KANG_KS45 | -24.371513 | 22.883032 | 1978 | FALCONBRIDGE |
| KANG_KS5 | -24.384000 | 22.874732 | 1978 | FALCONBRIDGE |
| KANG_KS54 | -24.367349 | 22.833028 | 1978 | FALCONBRIDGE |
| KANG_KS56 | -24.508966 | 22.874735 | 1978 | FALCONBRIDGE |
| KANG_KW1 | -24.400683 | 22.649721 | 1978 | FALCONBRIDGE |
| MOLOPO_M1 | -26.204722 | 22.174709 | 1978 | FALCONBRIDGE |
| MOLOPO_M11 | -26.167270 | 22.262206 | 1978 | FALCONBRIDGE |
| MOLOPO_M11A | -26.179708 | 22.270510 | 1978 | FALCONBRIDGE |
| MOLOPO_M16 | -26.304714 | 22.362215 | 1978 | FALCONBRIDGE |
| MOLOPO_M17 | -26.267264 | 22.308007 | 1978 | FALCONBRIDGE |
| MOLOPO_M18 | -26.254738 | 22.308012 | 1978 | FALCONBRIDGE |
| MOLOPO_M2 | -26.229765 | 22.212206 | 1978 | FALCONBRIDGE |
| MOLOPO_M20 | -26.300512 | 22.299708 | 1978 | FALCONBRIDGE |
| MOLOPO_M21 | -26.296334 | 22.353911 | 1978 | FALCONBRIDGE |
| MOLOPO_M22 | -26.242232 | 22.162207 | 1978 | FALCONBRIDGE |
| MOLOPO_M27 | -26.258820 | 22.170512 | 1978 | FALCONBRIDGE |
| MOLOPO_M3 | -26.267229 | 22.195506 | 1978 | FALCONBRIDGE |
| MOLOPO_M31 | -26.263035 | 22.124710 | 1978 | FALCONBRIDGE |
| MOLOPO_M36 | -26.217237 | 22.424712 | 1978 | FALCONBRIDGE |
| MOLOPO_M38 | -26.225567 | 22.133000 | 1978 | FALCONBRIDGE |
| MOLOPO_M4 | -26.171376 | 22.166410 | 1978 | FALCONBRIDGE |
| MOLOPO_M41 | -26.279708 | 22.341411 | 1978 | FALCONBRIDGE |
| MOLOPO_M43 | -26.317222 | 22.224704 | 1978 | FALCONBRIDGE |
| MOLOPO_M57 | -26.304755 | 22.324711 | 1978 | FALCONBRIDGE |
| MOLOPO_M8 | -26.238019 | 22.295515 | 1978 | FALCONBRIDGE |
| TSABONG_T13 | -26.114481 | 22.170505 | 1978 | FALCONBRIDGE |
| TSABONG_T34 | -26.100592 | 22.199707 | 1978 | FALCONBRIDGE |
| TSABONG_T45 | -26.117220 | 22.203906 | 1978 | FALCONBRIDGE |
| TSABONG_T5 | -26.125516 | 22.166407 | 1978 | FALCONBRIDGE |
| TSABONG_T6 | -26.142262 | 22.158010 | 1978 | FALCONBRIDGE |

| Kimberlite Name | Latitude | Longitude | Date Discovered | Prospecting Company |
|---|---|---|---|---|
| TSABONG_T7 | -26.146335 | 22.137207 | 1978 | FALCONBRIDGE |
| TSABONG_T9 | -26.088022 | 22.158004 | 1978 | FALCONBRIDGE |
| TSABONG_TN13 | -25.963043 | 22.437211 | 1978 | FALCONBRIDGE |
| TSABONG_TN14 | -25.971359 | 22.458017 | 1978 | FALCONBRIDGE |
| TSABONG_TN15 | -26.004729 | 22.516422 | 1978 | FALCONBRIDGE |
| TSABONG_TN16 | -25.983829 | 22.470519 | 1978 | FALCONBRIDGE |
| TSABONG_TN18 | -25.993691 | 22.491910 | 1978 | FALCONBRIDGE |
| JWANENG-10 | -24.646751 | 24.423678 | 1980 | DEBOT |
| KHUTSE-01 | -23.304917 | 24.390577 | 1981 | FALCONBRIDGE |
| KHUTSE-02 | -23.289664 | 24.358374 | 1981 | FALCONBRIDGE |
| KHUTSE-03 | -23.304019 | 24.390878 | 1981 | FALCONBRIDGE |
| KHUTSE-04 | -23.272928 | 24.240877 | 1981 | FALCONBRIDGE |
| KHUTSE-05 | -23.233271 | 24.163072 | 1981 | FALCONBRIDGE |
| KHUTSE-06 | -23.267440 | 24.221471 | 1981 | FALCONBRIDGE |
| KHUTSE-07 | -23.266535 | 24.236671 | 1981 | FALCONBRIDGE |
| KHUTSE-08 | -23.296004 | 24.370376 | 1981 | FALCONBRIDGE |
| KIKAO-01 | -23.088255 | 24.070571 | 1981 | FALCONBRIDGE |
| KIKAO-02 | -23.217430 | 24.140566 | 1981 | FALCONBRIDGE |
| KIKAO-03 | -23.154664 | 24.050369 | 1981 | FALCONBRIDGE |
| KIKAO-04 | -23.300718 | 24.066472 | 1981 | FALCONBRIDGE |
| KIKAO-05 | -23.142450 | 24.024765 | 1981 | FALCONBRIDGE |
| KIKAO-06 | -23.218584 | 24.149467 | 1981 | FALCONBRIDGE |
| KIKAO-07 | -23.216893 | 24.135871 | 1981 | FALCONBRIDGE |
| KIKAO-08 | -23.219687 | 24.145567 | 1981 | FALCONBRIDGE |
| Go25 | -22.626002 | 24.767292 | 1982 | FALCONBRIDGE |
| KANG_FE-01 | -24.042420 | 22.399722 | 1983 | DEBOT |
| KANG_FE-02 | -24.046512 | 22.403913 | 1983 | DEBOT |
| KGALAGADI-06 | -24.940897 | 22.184109 | 1983 | DEBOT |
| KGALAGADI-07 | -24.931397 | 22.216408 | 1983 | DEBOT |
| KGALAGADI-08 | -24.929791 | 22.216915 | 1984 | DEBOT |
| KHEKE-01 | -24.992323 | 23.003535 | 1985 | DEBOT |
| Go504 | -22.586355 | 24.615886 | 1985 | FALCONBRIDGE |
| Go130 | -22.590563 | 24.625682 | 1985 | FALCONBRIDGE |
| Go137 | -22.572832 | 24.691384 | 1986 | FALCONBRIDGE |
| Go136 | -22.586752 | 24.717783 | 1986 | FALCONBRIDGE |
| QUOXO_RIVER-01 | -22.844898 | 24.158071 | 1987 | DEBOT |
| QUOXO_RIVER-02 | -22.845645 | 24.174670 | 1987 | DEBOT |

# Appendix

| Kimberlite Name | Latitude | Longitude | Date Discovered | Prospecting Company |
|---|---|---|---|---|
| QUOXO_RIVER-03 | -22.537200 | 24.107273 | 1987 | DEBOT |
| QUOXO_RIVER-04 | -22.864108 | 24.239175 | 1987 | DEBOT |
| KGALAGADI-11 | -24.935025 | 22.186413 | 1987 | DEBOT |
| JWANENG-11 | -24.636747 | 24.393075 | 1990 | DEBOT |
| MARTINS_DRIFT-01 | -22.645531 | 28.114579 | 1990 | DEBOT |
| MARTINS_DRIFT-02 | -22.743016 | 27.862674 | 1990 | DEBOT |
| MARTINS_DRIFT-03 | -22.704061 | 27.883479 | 1990 | DEBOT |
| MARTINS_DRIFT-04 | -22.702988 | 27.897677 | 1990 | DEBOT |
| MARTINS_DRIFT-05 | -22.694430 | 27.893474 | 1990 | DEBOT |
| MARTINS_DRIFT-06 | -22.686604 | 27.887573 | 1990 | DEBOT |
| MARTINS_DRIFT-07 | -22.651924 | 28.077981 | 1990 | DEBOT |
| MARTINS_DRIFT-08 | -22.651015 | 28.078782 | 1990 | DEBOT |
| OKWA-01 | -22.595502 | 21.518799 | 1990 | DEBOT |
| OKWA-02 | -22.627689 | 21.556892 | 1990 | DEBOT |
| OKWA-03 | -22.633923 | 21.562796 | 1990 | DEBOT |
| OKWA-04 | -22.637654 | 21.490292 | 1990 | DEBOT |
| OKWA-05 | -22.570011 | 21.489696 | 1990 | DEBOT |
| OKWA-06 | -22.621594 | 21.536599 | 1990 | DEBOT |
| OKWA-07 | -22.643446 | 21.418492 | 1990 | DEBOT |
| ORAPA_BK18 | -21.491009 | 25.874621 | 1990 | DEBOT |
| ORAPA_DK3 | -21.500949 | 25.875018 | 1990 | DEBOT |
| ORAPA_BK19 | -21.220205 | 25.634419 | 1992 | DEBOT |
| ORAPA_BK20 | -21.226468 | 25.638119 | 1992 | DEBOT |
| ORAPA_BK21 | -21.223099 | 25.634816 | 1992 | DEBOT |
| ORAPA_BK22 | -21.241664 | 25.640619 | 1992 | DEBOT |
| ORAPA_BK23 | -21.240935 | 25.639817 | 1992 | DEBOT |
| ORAPA_BK24 | -21.254883 | 25.622015 | 1993 | DEBOT |
| ORAPA_BK25 | -21.245658 | 25.643011 | 1993 | DEBOT |
| ORAPA_BK26 | -21.245203 | 25.642610 | 1993 | DEBOT |
| ORAPA_BK27 | -21.346997 | 25.543912 | 1993 | DEBOT |
| ORAPA_BK28 | -21.330220 | 25.557315 | 1993 | DEBOT |
| ORAPA_BK29 | -21.421513 | 25.574614 | 1994 | DEBOT |
| ORAPA_BK30 | -21.383608 | 25.557515 | 1994 | DEBOT |
| ORAPA_BK31 | -21.454140 | 25.653113 | 1994 | DEBOT |
| ORAPA_BK32 | -21.346688 | 25.560509 | 1994 | DEBOT |
| ORAPA_BK33 | -21.380194 | 25.591613 | 1994 | DEBOT |
| ORAPA_BK34 | -21.305852 | 25.581112 | 1994 | DEBOT |

| Kimberlite Name | Latitude | Longitude | Date Discovered | Prospecting Company |
|---|---|---|---|---|
| ORAPA_BK35 | -21.494068 | 25.722117 | 1994 | DEBOT |
| ORAPA_BK36 | -21.262112 | 25.547013 | 1994 | DEBOT |
| JWANENG-12 | -24.619256 | 24.764180 | 1994 | DEBOT |
| JWANENG-13 | -24.635373 | 24.449772 | 1994 | DEBOT |
| JWANENG-14 | -24.637810 | 24.444479 | 1994 | DEBOT |
| ORAPA_BK37 | -21.319850 | 25.570016 | 1995 | DEBOT |
| ORAPA_BK38 | -21.285586 | 25.723821 | 1995 | DEBOT |
| ORAPA_DK4 | -21.503569 | 25.875133 | 1995 | REDAURUM |
| OKWA-08 | -22.603834 | 21.436687 | 1995 | DEBOT |
| ORAPA_BK39 | -21.337126 | 25.594117 | 1995 | DEBOT |
| ORAPA_BK40 | -21.267705 | 25.546110 | 1995 | DEBOT |
| ORAPA_BK41 | -21.303659 | 25.535915 | 1995 | DEBOT |
| ORAPA_BK42 | -21.331107 | 25.587310 | 1995 | DEBOT |
| ORAPA_BK43 | -21.334901 | 25.576513 | 1995 | DEBOT |
| KANG_KW6 | -24.483587 | 22.510335 | 1995 | SAMC |
| KANG_KW2 | -24.388978 | 22.640821 | 1996 | AFRIORE |
| KANG_KW3 | -24.439966 | 22.636526 | 1996 | AFRIORE |
| KANG_KW4 | -24.410992 | 22.647723 | 1996 | AFRORE |
| KANG_KW5 | -24.412965 | 22.653621 | 1996 | AFRIORE |
| ORAPA_AK15 | -21.306934 | 25.362107 | 1996 | DEBSWANA |
| ORAPA_BK44 | -21.395242 | 25.534513 | 1996 | DEBOT |
| ORAPA_BK45 | -21.340364 | 25.581609 | 1996 | DEBOT |
| ORAPA_BK46 | -21.398875 | 25.536811 | 1996 | DEBOT |
| ORAPA_BK47 | -21.327899 | 25.571114 | 1996 | DEBOT |
| ORAPA_BK48 | -21.321686 | 25.552308 | 1996 | DEBOT |
| ORAPA_AK16 | -21.402715 | 25.422405 | 1996 | DEBOT |
| ORAPA_AK17 | -21.433991 | 25.331607 | 1996 | DEBOT |
| ORAPA_AK18 | -21.432518 | 25.365408 | 1996 | DEBOT |
| ORAPA_AK19 | -21.422982 | 25.350909 | 1996 | DEBOT |
| ORAPA_BK49 | -21.273656 | 25.609109 | 1996 | DEBOT |
| ORAPA_BK50 | -21.394881 | 25.576512 | 1996 | DEBOT |
| ORAPA_BK51 | -21.332612 | 25.541316 | 1996 | DEBOT |
| SIKWANE-01 | -24.501636 | 26.443433 | 1996 | DEBOT |
| ORAPA_AK20 | -21.329075 | 25.410304 | 1997 | DEBOT |
| SIKWANE-02 | -24.500123 | 26.449530 | 1997 | DEBOT |
| SIKWANE-03 | -24.499883 | 26.458137 | 1997 | DEBOT |
| SIKWANE-04 | -24.501400 | 26.405336 | 1997 | DEBOT |

# Appendix

| Kimberlite Name | Latitude | Longitude | Date Discovered | Prospecting Company |
|---|---|---|---|---|
| SIKWANE-05 | -24.517156 | 26.415834 | 1997 | DEBOT |
| SIKWANE-06 | -24.510448 | 26.409528 | 1997 | DEBOT |
| SIKWANE-08 | -24.503130 | 26.408930 | 1997 | DEBOT |
| SIKWANE-09 | -24.508422 | 26.399736 | 1997 | DEBOT |
| GP194 | -22.549582 | 24.565878 | 1997 | TNK |
| GP173N | -23.085711 | 24.193799 | 1997 | TNK |
| GP173S | -23.095456 | 24.193636 | 1997 | TNK RESOURCES |
| GP211 | -23.101305 | 24.192626 | 1997 | TNK RESOURCES |
| GP234 | -23.102884 | 24.195140 | 1997 | TNK RESOURCES |
| ORAPA_BK52 | -21.300126 | 25.524508 | 1997 | DEBOT |
| SIKWANE-07 | -24.499960 | 26.454534 | 1997 | DEBOT |
| NXAUNXAU-06 THC06 | -18.958676 | 21.252899 | 1998 | REUNION MINING |
| NXAUNXAU-07 THC07 | -18.989978 | 21.240599 | 1998 | REUNION MINING |
| NXAUNXAU-09 THC09 | -18.957756 | 21.157590 | 1998 | REUNION MINING |
| NXAUNXAU-11 THC11 | -18.915061 | 21.147996 | 1998 | REUNION MINING |
| NXAUNXAU-12 THC12 | -18.905182 | 21.148690 | 1998 | REUNION MINING |
| NXAUNXAU-05 THC13 | -18.906109 | 21.158096 | 1998 | REUNION MINING |
| NXAUNXAU-10 THC14 | -18.937075 | 21.107191 | 1998 | REUNION MINING |
| NXAUNXAU-14 THC15 | -18.880404 | 21.048893 | 1998 | REUNION MINING |
| NXAUNXAU-04 THC16 | -18.902159 | 21.230191 | 1998 | REUNION MINING |
| NXAUNXAU-13 THC17 | -18.907557 | 21.227197 | 1998 | REUNION MINING |
| NXAUNXAU-03 THC18 | -18.918570 | 21.231093 | 1998 | REUNION MINING |
| NXAUNXAU-01 THC19 | -18.845783 | 21.349293 | 1998 | REUNION MINING |
| NXAUNXAU-15 THC42 | -18.830834 | 21.085994 | 1998 | REUNION MINING |
| NXAUNXAU-16 THC23 | -18.793784 | 21.095392 | 1998 | REUNION MINING |
| MOSOMANE-01 | -24.106960 | 25.896115 | 1998 | SAMC |
| MOSOMANE-02 | -24.107818 | 25.877621 | 1998 | SAMC |
| NXAUNXAU-08 THC39 | -18.947676 | 21.206194 | 1998 | REUNION MINING |
| NXAUNXAU-02 THC40 | -18.921597 | 21.238697 | 1998 | REUNION MINING |
| NXAUNXAU-17 THC41 | -18.820470 | 21.119588 | 1998 | REUNION MINING |
| THANKANE-01 | -24.812157 | 23.996561 | 2000 | DEBOT |
| THANKANE-02 | -24.907124 | 23.761673 | 2000 | DEBOT |
| KGARE-01 | -24.346080 | 24.882686 | 2000 | DEBOT |
| BOBONONG-01 | -21.641844 | 28.313914 | 2000 | KUKAMA EXPLORATION |
| DITSHEGWANE-01 | -24.197963 | 24.910996 | 2001 | DEBOT |
| JWANENG-15 | -24.730219 | 24.355269 | 2002 | DEBOT |
| DITSHEGWANE-02 | -24.218810 | 24.930285 | 2002 | DEBOT |

| Kimberlite Name | Latitude | Longitude | Date Discovered | Prospecting Company |
|---|---|---|---|---|
| DITSHEGWANE-03 | -24.180462 | 24.912009 | 2002 | DEBOT |
| ORAPA_BK55 | -21.483974 | 25.515183 | 2003 | BOTETI JV |
| DITSHEGWANE-04 | -24.238119 | 24.892534 | 2004 | DEBOT |
| KGARE-02 | -24.388256 | 24.729277 | 2004 | DEBOT |
| KGARE-03 | -24.379561 | 24.743575 | 2004 | DEBOT |
| ORAPA | -21.436556 | 25.371028 | 2004 | DEBOT |
| ORAPA | -21.452690 | 25.385147 | 2004 | DEBOT |
| ORAPA | -21.454406 | 25.387878 | 2004 | DEBOT |
| - | -21.397499 | 25.411301 | 2005 | DEBOT |
| - | -24.174604 | 24.814698 | 2005 | DEBOT |
| - | -24.302053 | 24.943165 | 2005 | DEBOT |
| - | -24.315004 | 24.826276 | 2005 | DEBOT |
| - | -24.313200 | 24.813499 | 2005 | DEBOT |
| - | -24.179399 | 24.909401 | 2005 | DEBOT |
| - | -24.372200 | 24.841801 | 2005 | DEBOT |
| MK81 | -26.186064 | 22.338513 | 2006 | Firestone Diamond |
| MK78 | -26.218123 | 22.396362 | 2007 | Firestone Diamond |
| MK82 | -26.218347 | 22.401219 | 2007 | Firestone Diamond |
| MK83 | -26.153984 | 22.486058 | 2007 | Firestone Diamond |
| MK79 | -26.165629 | 22.510455 | 2007 | Firestone Diamond |
| MK80 | -26.154853 | 22.473594 | 2007 | Firestone Diamond |
| MK71 | -25.959201 | 22.347421 | 2007 | Firestone Diamond |
| MK72 | -26.224172 | 22.338438 | 2007 | Firestone Diamond |
| MK73 | -26.036562 | 22.252364 | 2007 | Firestone Diamond |
| MK74 | -26.038772 | 22.248370 | 2007 | Firestone Diamond |
| MK75 | -26.208852 | 22.401865 | 2007 | Firestone Diamond |
| MK76 | -25.949095 | 22.332127 | 2007 | Firestone Diamond |
| MK77 | -25.935267 | 22.318931 | 2007 | Firestone Diamond |
| MAUN-01 | -20.500978 | 23.249750 | | AARL (FRANS BOSHOFF) |

**Note:** This data was supplied by the Botswana Dept. of Geological Survey in November, 2011. The table does not include kimberlite discoveries made post 2007, however the key in the kimberlite fields map in chapter 3 does contain all kimberlites known in 2011 - the additional kimberlites occur in Firestone Diamonds licences in the Tshabong and Kokong Kimberlite Fields.

# References


| | |
|---|---|
| Bedia, 2010 | Best of Botswana. |
| Boonen, S & Heens, C, undated | Diamond – a very special story, Mozaiek, in association with the Antwerp Diamond High Council. |
| Botswana Diamond Hub, undated | Botswana Diamond Hub Portfolio. |
| Botswana Dept. Surveys & Mapping, 2001 | Botswana National Atlas. |
| Capital Resources, 2011 | Botswana Resource Sector Overview and Botswana Country Economic Overview. |
| Debswana 1970-2010 | Annual Reports. |
| Debswana 2007 | Celebrating 25 years of sparkle at Debswana Jwaneng Mine. |
| Debswana, 2010 | The Debswana Diamond Book – Delivering the Diamond Dream, internal publication. |
| Debswana, 2012 | Jwaneng Cut 8 Project. |
| Daniels, L, 2004 | The history of Diamond Mining in Botswana, in Rough Diamond Review, September, 2004. |
| De Beers Botswana Mining Company (Pty) Ltd, 1982 | 1982 Annual Report. |
| De Beers, undated | Diamond Geology – published on www.debeersgroup.com |
| De Beers, 2008 | Pipeline – from kimberlite to Diamond Dream, Issue 1, 2008. |

| | |
|---|---|
| De Wit, M, 2010 | Identification of Global Diamond Metallogenic Clusters to assist exploration, in the Southern African Institute of Mining and Metallurgy, Diamonds source to use, 2010. |
| De Wit, M, Kostlin, E & Liddle, R, 2011 | Prospecting in Africa. |
| DTC, 2011 | Southern African Shining Light Awards Diamond Design Collection, 2010 - 2012. |
| Du Plessis, L and Sewawa, M, 2003 | FISH—State-of-the-art technology in final diamond recovery in the Journal of The South African Institute of Mining and Metallurgy, November, 2003. |
| Epstein, E.J, 1982 | The rise and fall of diamonds. |
| Even - Zohar, C, 2007 | From Mine to Mistress. |
| Field, M and Scott Smith B, 1998 | 1998 Textural and genetic classification schemes of kimberlites: a new perspective. Seventh International Kimberlite Conference, Cape Town, South Africa, 1998. |
| Field, M, Stiefenhofer J, Robey J, Kurszlaukis S, 2008 | Kimberlite - hosted diamond deposits of Southern Africa. |
| Field, M, Gibson, J. G, Wilkes, T.A Gababotse, J and P. Khutjwe, 1997 | The Geology of the Orapa A/K1 Kimberlite, Botswana: Further insight into the emplacement of kimberlite pipes In Russian Geology and Geophysics Vol. 38, No. I, pp. 24-19, 1997. |
| Firestone Diamonds, 2011 | Annual Report and Accounts, 2011. |
| Hamilton, W, Woolley, A & Bishop, A, 1975 | The Hamlyn Guide to Minerals, Rocks and Fossils. |

# References

| | |
|---|---|
| Kedikilwe, P.H.K, 2011 | Botswana Diamonds, 2020 and beyond. Presentation at Botswana Diamond Processing Conference, Gaborone, November, 2011. |
| Kjarsgaard, B.A. 1996 | Kimberlite-hosted Diamond; in Geology of Canadian Mineral Deposit Types, O.R. Eckstrand, W.D. Sinclair and R.I. Thorpe, editors, Geological Survey of Canada, Geology of Canada, Number 8, pages 561-568. |
| Lock, N, 1985 | Kimberlite exploration in the Kalahari region of Southern Botswana with emphasis on the Jwaneng Kimberlite Province. Botswana Geological Survey Report No DGS-REP-3622-1. |
| Lucara, 2010 | Technical Report on the Boteti Kimberlite Project, Botswana, prepared by MSA Geoservices (Pty) Ltd on behalf of Lucara Diamond Corporation. |
| Mogae, F, 2006 | Diamonds and Development. The Botswana Story. Speech prepared for Center for Global Development. |
| Pretorius, D.A. 1978. | The contribution of the aeromagnetic interpretation to an assessment of the mineral potential of Botswana. In: Reconnaissance Aeromagnetic Survey of Botswana1975-1977: Final Interpretation. Botswana Geological Survey, A1-A63. |
| Vision 2016, Botswana Government, 2011 | The Vision 2016 for Botswana. |
| Wannernburgh, A & Johnson, P, 1990 | Diamond People. |
| Wilson, A.N, 1971 | International diamond annual - a review of the world's diamond industry and trade - Botswana - a geologist's hunch. |
| World Jewellery Confederation, 2010 | Diamond Book - Diamonds — Terminology & Classification. |
| World Report Inc., 2006 | World Report Botswana – Celebrating 40 years of Independence. |

# Glossary & Abbreviations

| | |
|---|---|
| **Archaean** | A geological era comprising rock formations older than 2500 million years. |
| **BAHF** | Bulk Acid Handling Facility. |
| **Batswana** | Plural for Motswana. |
| **BOPA** | Botswana Press Agency. |
| **Bort Diamond** | A low-quality diamond, in granular aggregate or small fragments, valuable only in crushed or powdered form, especially for industrial use as an abrasive. |
| **Carat** | Is the unit weight for diamonds and derives from the Greek word 'keration' meaning fruit of the carob. The Carob tree (Ceratoniasiliqua) is an evergreen tree, with an edible pod containing seeds, which is native to the Mediterranean region. The seeds of the Carob were used on precision scales as units of weight for small quantities of precious gemstones because of their uniform size. The weight of an average carob seed is 200 milligrams. The weight of one carat is precisely 200 milligrams, or 0.2 grams. |
| **CAC** | Central Acidising Centre. |
| **CARP** | Completely Automated Recovery Plant. |
| **Cluster/Field/Province** | A group of individual kimberlite pipes or dykes ranging in numbers from several to 70 plus. Generally the cluster has a diameter of approximately 40 km. Several clusters form a field, whereas several fields are commonly referred to as a province. |
| **CPHT** | Carats per hundred tonnes of kimberlite ore, used to evaluate a kimberlite deposit. |
| **Crater** | Bowl-shaped depression on the earth's surface. |
| **Craton** | Is an old and stable part of the continental lithosphere, generally found in the interiors of tectonic plates. They are characteristically composed of ancient crystalline basement rock, which may be covered by younger sedimentary rock. |

# Glossary & Abbreviations

| | |
|---|---|
| Diamantaire | A highly skilled craftsman or artisan responsible for cutting, polishing and transforming a rough stone into a finished diamond ready for setting. |
| Diatreme | Carrot-shaped volcanic vent that has formed by explosive action. |
| DEBSWANA | Debswana Diamond Company. |
| DEF | Diamond Empowerment Fund. |
| DMS | Dense Media Separation. |
| DTP | Diamond Technology Park. |
| DTCB | Diamond Trading Company Botswana. |
| Facies | Is the sum of lithologic characteristics of a kimberlite deposit in a given place. |
| FISH | Fully Integrated Sort House. |
| Gabbro | A plutonic rock consisting mainly of Ca-plagioclase and clinopyroxene. An extrusive equivalent of basalt. |
| GIA | Gemological Institute of America. |
| Gneiss | A high grade metamorphosed granite with a course texture of mainly biotite and plagioclase. |
| Hypabyssal | Applied to medium-grained, intrusive igneous rocks which have crystallized at a shallow depth below the Earth's surface. |
| IDMA | International Diamond Manufacturers Association. |
| Kimberlite | A type of potassic volcanic rock best known for sometimes containing diamonds. It is named after the town of Kimberley in South Africa. |
| Lamproite | Are ultra-potassic mantle-derived volcanic and sub-volcanic rocks, which rarely contain diamonds. |
| Lithosphere | Solid outer portion of the earth including the crust and the portion of the upper mantle above the asthenosphere. Consists primarily of peridotite up to 175 km depth. |

| | |
|---|---|
| Macro-diamond | A diamond greater than 0.5mm in size. |
| Mantle | Zone of the earth below the crust and above the core (up to 3480 km). It is divided into the transition zone (up to 1000 km), the upper and lower mantle. |
| Metamorphic rock | A rock which has undergone physicochemical, mineralogical and structural changes when subjected to high temperature and pressure at depth within the earth's crust. |
| Micro-diamond | A diamond less than 0.5mm in size. |
| Motswana | Citizen of Botswana. |
| MTP | Main Treatment Plant. |
| Phanerozoic | This comprises rocks which are between 500 million years old and those of the present day. |
| Proterozoic | Comprises rock formations that are between 500 million years old and 2500 million years in age. |
| PULA | The official currency of the Rebuplic of Botswana – meaning "rain" in the national language of Setswana. |
| Sightholders | Are diamond cutting and polishing companies that have been licenced by the Government of Botswana to carry out cutting and polishing activities in the country. |
| SSV | Standard Selling Values. |
| Tuffisitic | Formed through fluidization processes. |
| Xenocryst | A crystal foreign to the host magma and picked up from the wall rocks en route from melting to crystallization site. Xenocrysts may be derived from the mantle, or shallow depths. |
| Xenolith | A foreign inclusion in an igneous rock. |
| WFDB | World Federation of Diamond Bourses. |

# Index

# Acknowledgements


| | |
|---|---|
| **Anja Assmann** | Principal Technical Education Officer, Ministry of Education and Skills Development - for supplying details of the syllabus for the new Jewellery design courses in Botswana. |
| **Marlaine Botha** | De Beers Consolidated Mines – for supplying old Botswana prospecting photos, diamond images and permission to use Gavin Lamont's obituary note |
| **Mike Brennan, Niall Young** | For supplying information and photos of Lerala Diamond Mine – Mantle Diamonds. |
| **Botswana National Archives** | For supplying archived diamond information and photos & Records |
| **Tony Cessford** | Acknowledged for his old Botswana prospecting images dated 1962 |
| **Rudie De Wet** | De Wet Drilling – for supplying photos of drilling |
| **De Beers Consolidated Mines** | For supplying photos of old prospecting and mining operations in Botswana |
| **DEBSWANA** | For supplying numerous images of all Debswana operations |
| **Lenayang Dimbungu** | Debswana – for supplying mining and mine planning information and images |
| **DTC / DTCB** | For supplying numerous images of diamonds etc. from DTC and DTCB |
| **Ribson Gabonowe** | Boteti Diamonds – for supplying photos of A/K 6 Karowe Mine |
| **Jim Gibson** | For supplying photos and information on the Discovery of Orapa Mine |
| **Lebuile Israel** | Deputy Town Clerk, Gaborone City Council – for permission to take photos of the Gaborone Fire Station |
| **Fred Jansen** | Debswana - for supplying old Botswana prospecting photos |
| **Polite Kutjwe** | Firestone Diamonds – for supplying photos of B/K 11 Mine |
| **Rachel Mothibatsela, Sarah Jele, Naledi Dikgomo-Goulden, Matshidiso Kamona and Boikhutso Ncube** | Debswana PCA – for supplying old and current images of Debswana mining operations and other photographs |

| | |
|---|---|
| **Takao Kaneko** | Gemological Institute of America – for supplying information and photos |
| **Hilary Koodirile** | Oodi College of Applied Arts and Technology – for supplying Information |
| **Kim Lanny** | Shrenuj Botswana – for supplying photos from Shrenuj Botswana |
| **Elijah Legwaila** | Debswana – for supplying processing flow sheets |
| **Lesole Lekgoa & Keoagile Mathule** | Botswana Geological Survey – for supplying prospecting licence and kimberlite data |
| **Mervin Lifshitz** | For supplying information and photos from Teemane Manufacturing Botswana |
| **Susan Matlolela** | Diamond Hub Liaison Officer - for supplying information |
| **Howard Marsden** | Gem Diamonds – for supplying photos of Ghaghoo Mine |
| **Mmetla Masire** | Debswana – for advice on Table of Contents |
| **Petronella Matumo** | For information and images on Private Collection |
| **Nchidzi Mmolawa** | Mineral Affairs Division – MMEWR – for supplying mineral production data |
| **Kago Mmopi & McDonald Mosete** | DTCB – for supplying photos from DTCB |
| **Ronald Modikana** | Business Development Manager, Peo Venture Capital, for permission to use Botswana "Unearthed" Private Collection Jewellery images |
| **Sellinah Mogojwa** | Central Statistics Office – for supplying mineral production data |
| **Maleta Mogwe** | For liaison with the Office of The President |
| **Motebedi Motebedi** | Anglo Coal Botswana – for supplying contacts of old Batswana Prospectors |
| **Orelebe Monnaapula** | Botswana Press Agency Photo section – for supplying some of the photos in chapter 1 & 2 |
| **Rutang Moses** | SAFDICO – for supplying information about SAFDICO |
| **Charles Njanjari** | For another great book design |

# Acknowledgements


| | |
|---|---|
| **Moeteledi D. Obuseng** | Debswana – for in pit mining and diamond processing information and photos |
| **Pauline Paledi-Mokou** | Executive Director BDMA for provision of information and photos |
| **Boineelo Peter** | Ministry of Finance and Development Planning – for supplying diamond revenue data |
| **Johannes Rademan** | Debswana – for supplying large stone data |
| **Vincent Ramokate** | De Beers Holdings Botswana – for supplying contacts of old Batswana Prospectors |
| **Martin Roberts** | Debswana – for supplying photos of the Jwaneng Mine J-Rep Project |
| **Mike Roberts, Angela Webb-Mpelega & Chipo Mrara** | De Beers Holdings Botswana – for supplying old & modern day prospecting photos |
| **R. Sandford** | Debswana – for information and images on diamond processing |
| **Prescilla Sejoe** | For assistance with book sponsorship |
| **Tebogo Selaelo** | Botswana Diamond Office for provision of information on the diamond center. |
| **Esther Kanaimba-Senai** | For diligent and comprehensive editing |
| **Kfir Teichman** | Steinmetz Diamonds Botswana – for supplying numerous photos from Steinmetz Diamonds |
| **Fred Tego** | Water Surveys Botswana – for producing GIS maps in Chapter 1 & 3 |
| **Max Terwindt** | ABN-AMRO – for supplying banking information |
| **Jacob Thamage** | For provision of information on the Botswana Diamond Hub |
| **Tom Tweedy** | De Beers Consolidated Mines, for permission to use De Beers Gavin Lamont's obituary note |
| **Walter van den Steen** | For supplying information and images from Diamond Manufacturing Botswana |
| **Andreas von Johannides** | For discussions on diamond prospecting in Botswana |
| **Dhammika Wickramasingha** | Schonberg Jewellers – for supplying photos of their jewellery |

481